MW00452319

›rary

THE SCIENCE OF WAR

STRATEGIES, TACTICS, AND LOGISTICS

THE BRITANNICA GUIDE TO WAR

THE SCIENCE OF WAR

STRATEGIES, TACTICS, AND LOGISTICS

EDITED BY ROBERT CURLEY , MANAGER, SCIENCE AND TECHNOLOGY

IN ASSOCIATION WITH

Published in 2012 by Britannica Educational Publishing
(a trademark of Encyclopædia Britannica, Inc.)
in association with Rosen Educational Services, LLC
29 East 21st Street, New York, NY 10010.

Copyright © 2012 Encyclopædia Britannica, Inc. Britannica, Encyclopædia Britannica, and the Thistle logo are registered trademarks of Encyclopædia Britannica, Inc. All rights reserved.

Rosen Educational Services materials copyright © 2012 Rosen Educational Services, LLC. All rights reserved.

Distributed exclusively by Rosen Educational Services.
For a listing of additional Britannica Educational Publishing titles, call toll free (800) 237-9932.

First Edition

Britannica Educational Publishing
Michael I. Levy: Executive Editor
Marilyn L. Barton: Senior Coordinator, Production Control
Steven Bosco: Director, Editorial Technologies
Lisa S. Braucher: Senior Producer and Data Editor
Yvette Charboneau: Senior Copy Editor
Robert Curley: Manager, Science and Technology

Rosen Educational Services
Nick Croce: Editor
Nelson Sá: Art Director
Cindy Reiman: Photography Manager
Brian Garvey: Designer
Introduction by Catherine Vanderhoof

Library of Congress Cataloging-in-Publication Data

The science of war: strategies, tactics, and logistics/edited by Robert Curley.
p. cm.—(The Britannica guide to war)
"In association with Britannica Educational Publishing, Rosen Educational Services."
Includes bibliographical references and index.
ISBN 978-1-61530-663-3 (library binding)
1. Military art and science—History. 2. Strategy—History. 3. Tactics—History. 4. War. I. Curley, Robert, 1955-
U102.S39 2012
355.409—dc23

2011017599

Manufactured in the United States of America

On the cover: As part of their Advanced Individual Training, Air and Missile Defense Crewmember trainees practice on the Joint Fires Multipurpose Dome, a $3.5 million combat simulator. *Photo Courtesy of U.S. Army/Ms. Marie Berberea (TRADOC)*

pp. 1, 32, 68, 102, 127 Library of Congress Geography and Map Division

CONTENTS

97
99
107

147
149
156
AFNETOPS/CC Vision

INTRODUCTION

Today in the world of 24-hour news, we cannot help but be aware of the brutality and human cost of war. It is difficult to imagine the festive atmosphere of parades and cheering as local regiments marched off to fight in the American Civil War, or even in the early days of World War I. Yet in spite of our lack of illusion about the reality of war, the idea of war still holds an enduring fascination. Cable television has an entire channel devoted to military history. Popular video game series are built around scenarios of both real and imagined warfare. Tens of thousands of participants take part in battle recreations from medieval up through modern times. Surely a significant part of this interest is because we understand how the outcomes of individual battles have changed the course of history and how often those outcomes seem to hinge on critical decisions about strategy, tactics, or logistics. It is easy to imagine "what if" scenarios in which history might have turned out very differently. It is those three critical aspects of war that we will explore in this volume—the strategy of war, the tactics of war, and the logistics of warfare from early history to the present day.

War is generally understood as armed conflict between two opposing military forces, waged with the goal of achieving some political purpose, such as conquest, independence, or acquisition of territory. War has certainly been a part of human history from the beginning of organized societies, as evidenced from our earliest written records, as well as archaeological finds such as the terra-cotta army of more than 7,000 soldiers, chariots, and horses buried with the first Qin emperor of China. Many early wars, involving the city-states of Mesopotamia and Egypt up through the Qin in China and Alexander the Great in the Mediterranean world, were primarily wars of conquest, resulting in one strong city-state or kingdom achieving control over its neighbours and creating a powerful empire.

Over the course of history, we see how warfare becomes a more nuanced political tool, with a variety of motives. We also begin to find written histories and analyses of the strategy of war, which can be defined as the interaction of political, economic, and military activities to achieve the objects of war. One of the earliest volumes of military strategy is *The Art of War,* attributed to the Chinese philosopher Sunzi in approximately the 5th century BCE. At about the same time the Greek historian Thucydides was writing his *History of the Peloponnesian War,* with insights into the battle between Athens and Sparta for control over the

An American soldier wipes his eyes on October 19, 2010, in Kajaki, Afghanistan. His comrade has just been killed by an improvised explosive device (IED). Scott Olson/Getty Images

Greek world. Historians of military strategy still study and draw lessons from these ancient theorists, as well as from more modern writers, from Machiavelli in the 15th century to Carl von Clausewitz in the 19th century. Because strategy encompasses both the political and the military realm, it also reflects changes in social structures, communications, technology, and popular opinion. Theories and strategies for warfare continue to develop today in light of such changes as the possibility of nuclear annihilation on the one hand and the growth of small-scale guerrilla warfare and terrorism on the other.

If strategy is seen as the grand plan for war, tactics is the science of how battles are actually waged—how troops are organized, what weapons will be used, and the execution of the battle plan. While strategy typically takes place at the highest level of command or with the political leadership, tactics most frequently are the decisions of commanders in the field.

The study of the tactics of war is also inextricably linked to advances in the tools of war—the science of weaponry. In the earliest days of organized warfare, weapons were hand-held blades and cudgels, and warfare was by necessity a matter of close combat between individual combatants. By approximately 2000 BCE, the introduction of the chariot revolutionized warfare, allowing for greater mobility and speed and allowing the armies of the western steppes to overrun much of Europe and Asia. From chariot warfare evolved another new tactic—soldiers on horseback, or cavalry. These warriors could move swiftly, remaining out of range of foot soldiers while attacking with bows and other long-range weapons. In other cases cavalry were used in combination with traditional troops, overwhelming the opponent's foot soldiers with the advantage of height and strength that the horses provided. As these various types of different troops and weaponry became available to battle commanders, tactics also evolved to make the best use of their strengths and the enemy's weaknesses. By the time of Caesar's conquests, Roman legions became the most advanced fighting force yet developed, consisting of skillfully arranged ranks of warriors with a variety of weapons, well trained and organized in a carefully planned style of attack in battle.

Following the fall of the Roman Empire, warfare in the Middle Ages was generally on a much smaller scale, with armoured knights on horseback, sword-wielding infantry, and the introduction of armour-piercing crossbows and longbows. The next major advance in warfare was the introduction of gunpowder. Gunpowder not only allowed for individual muskets and other firearms, but also for the development of artillery—cannons capable of inflicting great damage at a distance. By the 16th century, all European armies had adopted an organizational structure consisting of these three categories of troops: infantry, cavalry, and artillery. Tactics for how to use these troops in the field also evolved accordingly, and drilling and training of professional armies once again took on greater importance, since success in battle was dependent on the coordinated deployment of the various troops.

A converse effect can also be seen. As weaponry became ever more sophisticated and increasingly deadly at long range, the battlefield tactic of massing large groups of troops against each other was quickly revealed as a suicidal approach to war. In the face of artillery and gunfire, troops quickly learned to spread out and take cover, rather than providing a single massive target for the opponent's attack. The ability of a single commander to control the movements and objectives of an entire army became much more problematic as well, leading most armies to rely on a much more decentralized approach to command, with more autonomy for smaller groups in individual actions against the enemy, acting within the larger battle plan as designed by the central command.

Another outcome and tactical change was the increasing benefit of taking up a protected defensive position and forcing the attacking troops into a more exposed position. However, by the time of World War I this tactic ended in a deadly stalemate with armies on both sides dug into defensive trenches, resulting in a war of attrition as each side attempted to gain a few hundred yards of territory. The lessons of this war had profound impacts on military strategy and tactics, as well as both technical and tactical innovations in the design and use of tanks and airplanes, both first employed during World War I.

The lessons to be learned from World War I's battle tactics differed among nations and military strategists, however. The saying that we always fight the last war is frequently applied to the French plan to defend their country against any future attack by building a massive defensive barrier known as the Maginot Line. That strategy proved woefully inadequate against the new tactics of the German army in World War II, however, which had taken a different lesson from World War I—that speed of attack, use of armoured tanks to overrun enemy lines, and outflanking the opponent rather than attacking head-on were the keys to victory in modern warfare. Ultimately, however, the rapid blitzkrieg style of attack and reliance on motorized tank convoys proved undone by the third aspect of the science of war: logistics.

Logistics is perhaps the least glamorous job in wartime, but it is the foundation on which all other aspects of battle are founded. Logistics is the science of transportation, communication, medical treatment, and supplies—all the things that an army requires in order to coordinate and fight its battles. A constant danger in battle campaigns is that the forward troops will outpace or be cut off from the available supply lines, leaving tanks and other vehicles with no source of fuel, troops with no reserves of ammunition or access to reinforcements, and even in some cases, such as the Germans' attack on Stalingrad, without necessary food and clothing.

By the end of World War II, technical and scientific advances had transformed almost every aspect of warfare, from the development of submarines under the oceans to intercontinental rockets attacking through the skies, to the splitting of the atom and the creation of the atomic bomb. It can be argued that it is only in the age of nuclear war that the prevention of war became an essential element of strategy, with the possibility that a full-fledged nuclear conflict might result in devastation so complete as to render victory for either side impossible to achieve. As the major military powers built up their arsenals of nuclear weaponry with the goal of making any attack against them unthinkable due to their retaliatory power, it was only among—or against—small, conventionally armed countries that war could be carried on more or less as before. And in those wars, increasingly the terms of battle shifted from conventional to guerrilla-style warfare.

The term *guerrilla* is a diminutive of the Spanish *guerra*, so it literally translates to "little war." It traditionally refers to the efforts of irregular, local partisan forces, fighting by means of small-scale, fast-moving attacks against either troops or supply lines of the opposition armies. Although the term came into use only in the 1800s, this style of combat has been employed in wars going back to the earliest battle campaigns. In fact, it could be said to date back even to pre-civilization, beginning with raiding parties between neighbouring tribes and villages. In more modern times, guerrilla forces played important roles in the American Revolution as well as in defeating Napoleon's ambitions of empire. Local guerrilla fighters frequently attempted, usually in vain, to prevent colonization of Africa and Asia in the 19th century. Resistance movements in occupied Europe and Asia during World War II used guerrilla tactics to sabotage German and Japanese supply lines and assist the Allied cause.

In the post–World War II period, virtually every war around the world can be classified as a guerrilla war to one degree or another. Mao Zedong's rise to power in China was fueled by the revolutionary fervour of his guerrilla army. First the French and then the Americans were defeated by local guerrilla forces in Vietnam. Russia fought and ultimately lost a costly guerrilla war in Afghanistan. Peace remains elusive in the Middle East with a variety of Palestinian and Arab organizations engaged in guerrilla style attacks against Israel, not necessarily sanctioned by any national government. In fact, many guerrilla movements are insurgency movements intended to overthrow existing regimes, rather than a battle between two national governments or larger groups of allied forces as in traditional warfare, and the boundaries between guerrilla warfare and terrorism become difficult to discern and maintain.

How can military strategy evolve to battle combatants such as al-Qaeda, a non-state-sponsored international terrorist movement that follows none of the traditional rules of war and makes no distinction between civilian and military targets? Or how can military tactics help us to understand and resolve ethnic conflicts within the population of a single nation, as have taken place in Bosnia, Serbia, Chechnya, and Indonesia, as well as so many nations of Africa? It is certainly true that many of the lessons of modern warfare no longer seem to apply. The wars of the 21st century seem to have more in common, in some ways, with the localized, often brutal conflicts of the European Middle Ages, but with frightening new advances in technology and weaponry. Politicians need to create new strategic paradigms for why and how to engage in wars, and commanders need to create new tactics for fighting them, including the new sciences of nation-building and counter insurgency efforts to win the hearts and minds of the local citizenry. Only history will tell if we have been wise enough to learn the right lessons and apply them to our 21st-century challenges.

CHAPTER 1

STRATEGY

Strategy is the science or art of employing all the military, economic, political, and other resources of a country to achieve the objects of war. The term *strategy* derives from the Greek *strategos*, an elected general in ancient Athens. The *strategoi* were mainly military leaders with combined political and military authority, which is the essence of strategy. Because strategy is about the relationship between means and ends, the term has applications well beyond war: it has been used with reference to business, the theory of games, and political campaigning, among other activities. It remains rooted, however, in war, and it is in the field of armed conflict that strategy assumes its most complex forms.

FUNDAMENTALS OF STRATEGY

Theoreticians distinguish three types of military activity: (1) tactics, or techniques for employing forces in an engagement (e.g., seizing a hill, sinking a ship, or attacking a target from the air), (2) operations, or the use of engagements in parallel or in sequence for larger purposes, which is sometimes called campaign planning, and (3) strategy, or the broad comprehensive harmonizing of operations with political purposes. Sometimes a fourth type is cited, known as grand strategy, which encompasses the coordination of all state policy, including economic and diplomatic tools of statecraft, to pursue some national or coalitional ends.

Strategic planning is rarely confined to a single strategist. In modern times, planning reflects the contributions of committees and working groups, and even in ancient times the war council was a perennial resort of anxious commanders. For example, the ancient Greek historian Thucydides' *History of the Peloponnesian War* (c. 404 BCE) contains marvelous renditions of speeches in which the leaders of different states attempt to persuade their listeners to follow a given course of action. Furthermore, strategy invariably rests on assumptions of many kinds—about what is lawful or moral, about what technology can achieve, about conditions of weather and geography—that are unstated or even subconscious. For these reasons, strategy in war differs greatly from strategy in a game such as chess. War is collective; strategy rarely emerges from a single conscious decision as opposed to many smaller decisions; and war is, above all, a deeply uncertain endeavour dominated by unanticipated events and by assumptions that all too frequently prove false.

Such, at least, has been primarily the view articulated by the greatest of all Western military theoreticians, the Prussian general Carl von Clausewitz. In his classic strategic treatise, *On War* (1832), Clausewitz emphasizes the uncertainty under which all generals and statesmen labour (known as the "fog of war") and the tendency for any plan, no matter how simple, to go awry (known as "friction"). Periodically, to be sure, there have been geniuses who could steer a war from beginning to end, but in most cases wars have been shaped by committees. And, as Clausewitz says in an introductory note to *On War*, "When it is not a question of acting oneself but of persuading others in discussion, the need is for clear ideas and the ability to show their connection with each other"—hence the discipline of strategic thought.

Clausewitz's central and most famous observation is that "war is a continuation of politics by other means." Of course war is produced by politics, though in common parlance war is typically ascribed to mindless evil, the wrath of God, or mere accident, rather than being a continuation of rational diplomacy. Moreover, Clausewitz's view of war is far more radical than a superficial reading of his dictum might suggest. If war is not a "mere act of policy" but "a true political instrument," political considerations may pervade all of war. If this is the case, then strategy, understood as the use of military means for political ends, expands to cover many fields. A seeming cliché is in fact a radical statement.

There have been other views, of course. In *The Art of War*, often attributed to Sunzi (5th century BCE) but most likely composed during a tempestuous time called the Warring States period between (475–221 BCE), war is treated as a serious means to serious ends, in which it is understood that shrewd strategists might target not an enemy's forces but intangible objects—the foremost of these being the opponent's strategy. Though this agrees with Clausewitz's ideas, *The Art of War* takes a very different line of argument in other respects. Having much greater confidence in the ability of a wise general to know himself and his enemy, *The Art of War* relies more heavily on the virtuosity of an adroit commander in the field, who may, and indeed should, disregard a ruler's commands in order to achieve war's object. Where *On War* asserts that talent for high command differs fundamentally from military leadership at lower levels, *The Art of War* does not seem to distinguish between operational and tactical ability; where *On War* accepts battle as the chief means of war and extensive loss of human life as its inevitable price, *The Art of War* considers the former largely avoidable ("the expert in using the military subdues the enemy's forces without going to battle") and the latter proof of poor generalship; where *On War* doubts that political and military leaders will ever have enough information upon which to base sound decisions, *The Art of War* begins and concludes with a study of intelligence collection and assessment.

To some extent, these approaches to strategy reflect cultural differences. Clausewitz is a product of a combination of the Enlightenment, an 18th-century period that emphasized reason, and early Romanticism, an early 19th-century philosophy that rebelled against pure rationality and emphasized imagination—and which showed a fascination with folk culture and national origins; *The Art of War* has its roots in Daoism, a Chinese religious-philosophical tradition that teaches unassertive action and simplicity and is concerned with obtaining long life and good fortune, often by magical means.

Historical circumstances explain some of the differences as well. Clausewitz laboured under the impact of 20 years of war that followed the French Revolution and the extraordinary personality of the famous French general Napoleon Bonaparte whose armies subjugated much of Europe in the periods between 1799–1814/15. As noted earlier, *The Art of War* was written during the turmoil of the Warring States period. There also are deeper differences in thinking about strategy that transcend time and place. In particular, differences in contemporary discussions of strategy persist between optimists, who think that the wisely instructed strategist has a better than even chance (other things being equal) to control his fate, and pessimists (such as Clausewitz), who believe that error, muddle, and uncertainty are the norm in war and therefore that chance plays a more substantial role. In addition, social scientists, exploring such topics as inadvertent war or escalation, have been driven by the hope of making strategy a rational and predictable endeavour. Historians, by and large, side with the pessimists: in the words of British historian Michael Howard, one of the best military historians of the 20th century, most armies get it wrong at the beginning of a war.

CAESAR CONQUERS GAUL—AND WINS ROME

In ancient Rome a nobleman won distinction for himself and his family by securing election to a series of public offices, and Gaius Julius Caesar, a scion of the patrician clan the Julii, set himself on this course at a young age. From the beginning, he probably aimed at winning office not just for the sake of the honours but also to achieve the power he needed to put the misgoverned Roman state into better order (according to his own ideas). In 62 BCE he was elected praetor, a one-year judicial position that frequently led to becoming a provincial governor—and thus gave ample opportunity for plunder. In due course Caesar obtained the governorship of a region called Farther Spain (which included parts of modern-day Spain and Portugal) for 61–60 BCE, and a military expedition beyond the northwest frontier of this province enabled him to win loot for himself as well as for his soldiers, with a balance left over for the treasury. This success enabled him, after his return to Rome in 60 BCE, to stand for and win the powerful consulship for 59.

As consul, Caesar was awarded the governorship of Cisalpine Gaul (northern Italy between the Alps, the Apennines, and the Adriatic) as well as Illyricum (on the Balkan Peninsula across the Adriatic). When the governor-designate of Transalpine Gaul suddenly died, this province (embracing all of what is now France and Belgium, along with parts of Germany, the Netherlands, and Switzerland) was also assigned to Caesar. Cisalpine Gaul gave Caesar a military recruiting ground; Transalpine Gaul gave him a springboard for conquests beyond Rome's northwest frontier.

In 58 BCE Rome's northwestern frontier ran from the Alps down the left bank of the upper Rhône River to the Pyrenees, skirting the southeastern foot of the Cévennes and including the upper basin of the Garonne River without reaching the Gallic shore of the Atlantic. Caesar, clad in the blood red cloak he usually wore "as his distinguishing mark of battle," intervened beyond this line in 58, first to drive back the Helvetii, who had been migrating westward from their home in what is now central Switzerland. In 57 Caesar subdued the distant and warlike Belgic group of Gallic peoples in the north.

In 56 the Veneti, in what is now southern Brittany, started a revolt in the northwest. Caesar reconquered the Veneti with some difficulty and treated them barbarously. In 55, he bridged the Rhine just below Koblenz to raid Germany on the other side of the river, and then crossed the Channel to raid Britain. In 54 he raided Britain again and subdued a serious revolt in northeastern Gaul. In 53 he subdued further revolts in Gaul and bridged the Rhine again for a second raid.

The crisis of Caesar's Gallic war came in 52. The peoples of central Gaul found a national leader in the Arvernian chieftain Vercingetorix. Vercingetorix used guerrilla warfare to harass Caesar's supply lines and cleverly offered to engage Caesar's forces on terrain unfavourable to the Romans. He successfully held the Arvernian hill-fort of Gergovia against an assault by Caesar. Vercingetorix followed up this victory by an attack on the Roman army, the failure of which compelled him to retreat with 80,000 troops to the prepared fortress of Alesia (in east-central France). Caesar, with a force of 60,000 men, laid siege to the fortress and was able to force its surrender after he had defeated the Gauls' reserve army in the field. Vercingetorix was taken to Rome in chains, exhibited in Caesar's triumph, and executed six years later.

During the winter of 52–51 and the campaigning season of 51, Caesar crushed a number of sporadic further revolts. He spent the year 50 BCE in organizing the newly conquered territory. After that, he was ready to settle his accounts with his opponents at home. On Jan. 10–11, 49 BCE, Caesar led his troops across the little river Rubicon, the boundary between Cisalpine Gaul and Italy proper. He thus violated the law (the Lex Cornelia Majestatis) that forbade a general to lead an army out of the province to which he was assigned. His act amounted to a declaration of war against the Roman Senate and resulted in a three-year civil war that left Caesar ruler of the Roman world.

HISTORICAL DEVELOPMENT

Throughout most of the centuries of warfare, military men have devised their own strategies and insisted they were the best. *The Art of War* was one of the earliest compilations on strategy. Sunzi's insistence on the political aspect of war was influential on later generals. Altogether, Sunzi set forth 13 principles of generalship. Much later Napoleon decided there were at least 115 maxims needed to guide generals. In the United States the Civil War general Nathan Bedford Forrest needed only one: get there first with the most men. He was in overall agreement with the Prussian Clausewitz, for whom defeat of the enemy's armed forces on the battlefield was the heart of strategy. Although generals have long disagreed, most principles of strategy include clarifying the objective of the campaign; unity of command; mass concentration of force; the effort to achieve surprise; proper movement of forces, their security from surprise attack, sabotage, or subversion; and simplicity of operation.

STRATEGY IN ANTIQUITY

The ancient world offers the student of strategy a rich field for inquiry. Indeed, the budding strategist is probably best advised to begin with Thucydides' *History of the Peloponnesian War*, which describes the contest between two coalitions of Greek city-states between 431 and 404 BCE. Athens, a predominantly maritime power, led the former members of the Delian League (a confederacy of ancient Greek states now incorporated in the Athenian empire) against the Peloponnesian League, which was led by Sparta, a cautious land power where every male citizen spent years in training as a soldier. In the opening speeches rendered by Thucydides, the two leaders, Pericles of Athens and

Archidamus II of Sparta, wrestle with strategic issues of transcendent interest: How shall they bring their strengths to bear on their enemy's weakness, particularly given the different forms of power in which the two coalitions excel? How will the nature of the two regimes—the volatility and enterprising spirit of democratic Athens, the conservatism of highly militarized Sparta—shape the contest?

From his study of the Peloponnesian War, the 19th-century German military historian Hans Delbrück drew a fundamental distinction between strategies based on

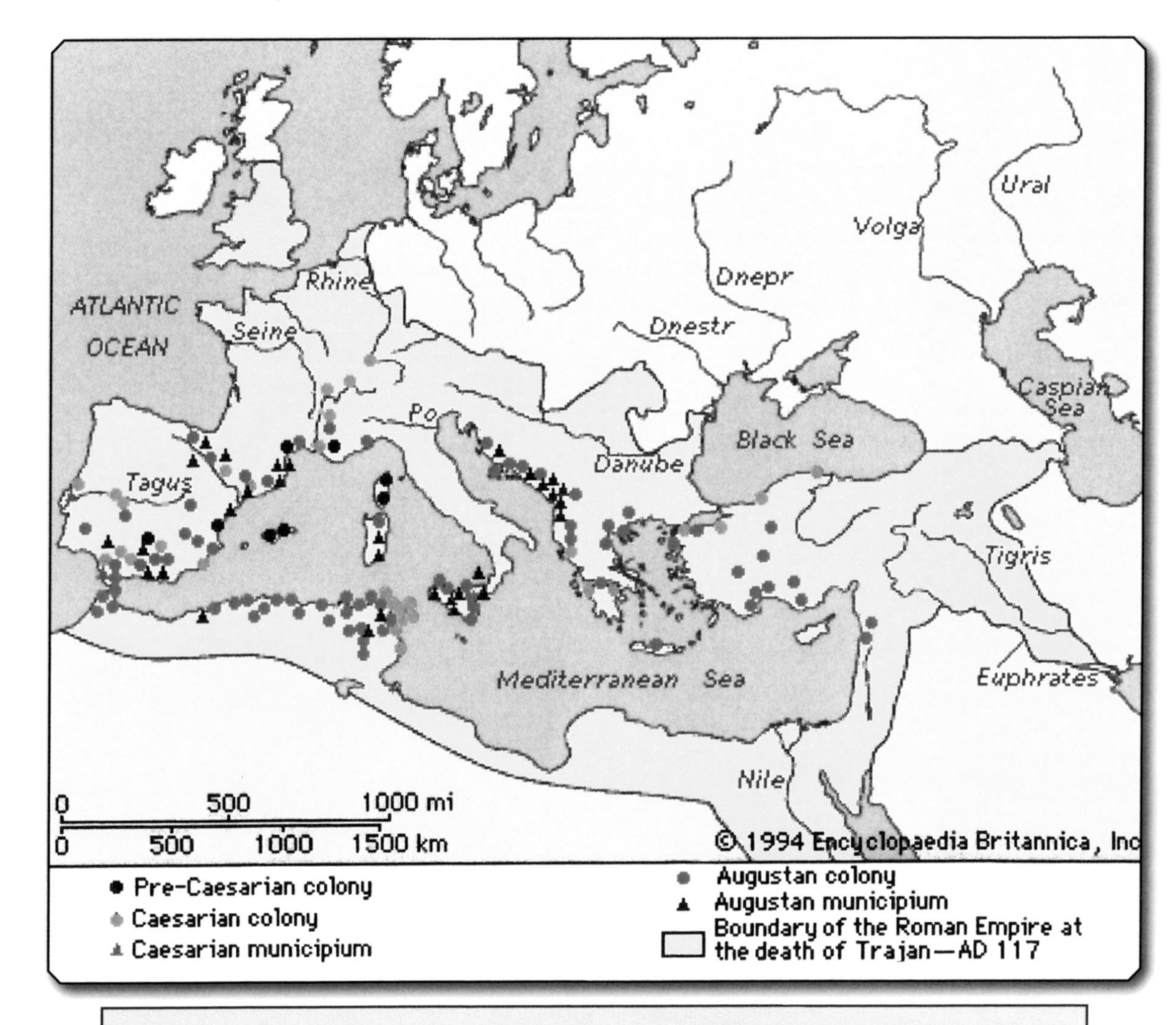

Over time, Rome's relentless drive for growth allowed it to become a mighty empire. From the time of its first emperor, Augustus, to Emperor Trajan, it grew to its largest extent, covering parts of three continents. It is shown here at the time of the death of Trajan in 117 CE.

overthrow of the opponent and those aimed at his exhaustion. Both Sparta and Athens pursued the latter strategy; the former was simply unavailable, given their fundamental differences as military powers. Although the Spartans eventually won, Greece was never the same.

But if ancient Greece is a story of distinctive city-states that shared a long, complex history of cooperation and competition, the rise of ancient Rome is far more a story of institutions. The story of Roman strategy seems one of a collective approach to war. Rome's great strength, historians argue, stemmed from political institutions that turned internal divisions into an engine of external expansion, that allowed for popular participation and executive decision, and that concentrated strategic decision making in a powerful Senate composed of the leading men of Rome. To its unique political constitution was added the Roman legion, a form of military organization far more flexible and disciplined than anything the world had yet seen—a fabulous tool for conquest and, in its attention to detail, from the initial selection of soldiers to their construction of camps to their rotation on the battle line, a model imitated in succeeding centuries.

Rome's conquest of the Mediterranean world illustrates the idea of a tacit or embedded strategy. Rome's ruthlessness in dividing its enemies, in creating patron-client relationships that would guarantee its intervention in more civil wars, its cleverness in siding with rebels or dissidents in foreign states, and its relentlessness in pursuing to annihilation its most serious enemies showed remarkable continuity throughout the Roman Republic and later, the Roman Empire.

The Second Punic War (218–201 BCE), one in a series of wars between the Roman Republic and the Carthaginian (Punic) Empire of northern Africa, illustrates these propositions well. There were two leading Roman figures of note throughout the war: Roman commander Quintus Fabius Maximus Verrucosus (Cunctator), who delayed and bought time while Rome recovered from its initial disastrous defeats (the nickname Cunctator means "delayer" in Latin), and General Scipio Africanus the Elder, who delivered the final blow of the Second Punic War to Carthage at the Battle of Zama (202 BCE). It does not appear that either was the equal of Hannibal, the brilliant Carthaginian general who administered defeat after defeat to superior Roman armies on their home turf. More important than personalities, however, was Rome's unflinching determination to pursue its enemies, quite literally to the death. Hannibal was cornered in the Bithynian village of Libyssa and committed suicide, following a demand from Rome that he be turned over by Antiochus III of Syria, whom he had aided in rebellion against Rome following the defeat of Carthage. And Carthage itself—long the target of the grim Roman senator Marcus Porcius Cato's insistence that it be destroyed (he famously took to ending every oration with the words "Ceterum censeo delendam esse Carthaginem," which translate as "Besides which, my opinion is that Carthage must be destroyed")—was wiped out of existence in the Third Punic War (149–146 BCE), which was provoked by Rome for the purpose of finishing off its most dangerous potential opponent.

Medieval Strategy

Most military histories skim over the Middle Ages (the period from the collapse of Roman civilization in the 5th century CE to the Renaissance—variously interpreted as beginning in the 13th, 14th, or 15th century, depending on the region of Europe—incorrectly believing it to be a period in which strategy was displaced by a combination of banditry and religious fanaticism. Certainly, the sources for medieval strategic thought lack the literary appeal of the classic histories of ancient Greece and Rome. Nevertheless, Europe's medieval period may be of especial relevance to the 21st century. In the Middle Ages there existed a wide variety of entities—from empires to embryonic states to independent cities to monastic orders and more—that brought different forms of military power to bear in pursuit of various aims. Unlike the power structures in the 18th and 19th centuries, military organizations, equipment, and techniques varied widely in the medieval period: the pikemen of Swiss villages were quite different from the mounted knights of western Europe, who in turn had little in common with the light cavalry soldiers of the Arabian heartland. The strategic predicament of the Medieval Byzantine Empire, which was centered in what is now the country of Turkey, poised between Europe and Asia—beset by enemies that ranged from the highly civilized Persian and Arab empires to marauding barbarians—required, and elicited, a complex strategic response, including a notable example of dependence on high technology. Greek fire, a liquid incendiary agent, was one tool that helped the embattled Byzantine Empire to beat off attacking fleets and preserve its existence until 1453.

In Delbrück's parlance, medieval warfare demonstrated both types of strategy—overthrow and exhaustion. The Crusader states of the Middle East—a series of Christian states established in what is now Israel and Lebanon by European Crusaders between the 11th and 14th centuries—were gradually exhausted and overwhelmed by constant raiding warfare and the weight of numbers of their Muslim opponents. On the other hand, one or two battles proved decisive, most notably the ruinous disaster at the Battle of Ḥaṭṭīn (1187), a battle in northern Palestine where 18,000 soldiers under the leadership of the powerful Muslim leader Saladin crushed a force of 15,000 Christian soldiers, dooming the Crusader kingdom of Jerusalem.

Medieval strategists made use of many forms of warfare, including set-piece battles, of course, as well as the petty warfare of raiding and harassment. But they also improved a third type of warfare—the siege, or, more properly, poliorcetics, the art of both fortification and siege warfare. Castles and fortified cities could eventually succumb to starvation or to an assault using battering rams, catapults, and mining (also known as sapping, a process in which tunnels are dug beneath fortification walls preparatory to using fire or explosives to collapse the structure), but progress in siege warfare was almost always slow and painful. On the whole, it was substantially easier to defend a fortified position than to attack one, and even a small force could achieve a disproportionate military advantage by occupying a defensible place. These facts, combined with the primitive public-health practices of many medieval armies, the poor condition of road networks, and the poverty of an agricultural system that did not generate much of a surplus upon which armies could feed, meant limits on the tempo of war and in some measure on its decisiveness as well—at least in Europe.

The story was different in East and Central Asia, particularly in China, a wealthy and civilized society in which the well-developed infrastructure of farms, roads, villages, and cities and the relatively open terrain made the Chinese easy prey for mobile cavalry units such as those of the Mongol invaders from the north, bent on pillage and conquest.

But it was in Europe that a competitive state system, fueled by religious and dynastic tensions and making use of developing civilian and military technologies, gave birth to strategy as it is known today.

Strategy in the Early Modern Period

The development of state structures, particularly in western Europe, during the 16th and 17th centuries gave birth to strategy in its modern form. "War makes the state, and the state makes war," in the words of American historian Charles Tilly. The development of centralized bureaucracies and, in parallel, the taming of independent aristocratic classes yielded ever more powerful armies and navies. As the system of statecraft gradually became secularized—witness the careful policy pursued by France under the influential cardinal Armand-Jean du Plessis, duc de Richelieu, chief minister to King Louis XIII from 1624 to 1642, which showed how strategy was becoming more subtle over time. He was willing to persecute powerful Huguenot Protestants at home—to Richelieu the Huguenots constituted a state within a state, with the civil government of major cities in their hands and considerable military force at their disposal. But because he also had reason to believe that Spain's Catholic Habsburg rulers had designs on siezing France, he supported Protestant powers abroad during the Thirty Years' War (1618–48). This series of wars was fought by various nations for various reasons, including religious, dynastic, territorial, and commercial rivalries between the Holy Roman Empire, which was Roman Catholic and Habsburg, and a network of Protestant towns and principalities that relied on the chief anti-Catholic powers of Sweden and the United Netherlands. In this as in many other ways, the early modern period witnessed a return to Classical roots. Even as drill masters studied ancient Roman textbooks to recover the discipline that made the legions formidable instruments of policy, so too did strategists return to a Classical world in which the logic of foreign policy shaped the conduct of war.

For a time, the invention of gunpowder and the development of the newly centralized state seemed to shatter the dominance of defenses: medieval castles could not withstand the battering of late 15th- or early 16th-century artillery. But the invention of carefully designed geometric fortifications (known as the *trace italienne*) restored much of the balance. A well-fortified city was once again a powerful obstacle to movement, one that would require a great deal of time and trouble to reduce. The construction of belts of fortified cities along a country's frontier was the keynote of strategists' peacetime conceptions.

Poliorcetics was no longer a haphazard art practiced with greater or lesser virtuosic skill but increasingly a science in which engineering and geometry played a central role; cities fell not to starvation but to methodical bombardment, mining, and, if necessary, assault. Armies also began to acquire the rudiments, at least, of modern logistical and health systems; though they were not quite composed of interchangeable units, they at least comprised a far more homogeneous and disciplined set of suborganizations than they had since Roman times. And, in a set of developments rarely noticed by military historians, the development of ancillary sciences, such as the construction of roads and highways and cartography, made the movement of military organizations not only easier but more predictable than ever before.

Strategy began to seem more like technique than art, science rather than craft. Practitioners, such as the 17th-century French engineer Sébastien Le Prestre de Vauban and the 18th-century French general and military historian Henri, baron de Jomini, began to make of war an affair of rules, principles, and even laws. Not surprisingly, these developments coincided with the emergence of military schools and an increasingly scientific and reforming bent—artillerists studied trigonometry, and officers studied military engineering. Military literature flourished: *Essai général de tactique* (1772), by Jacques Antoine Hippolyte, comte de Guibert, was but one of a number of thoughtful texts that systematized military thought, although Guibert (unusual for writers of his time) had inklings of larger changes in war lying ahead. War had become a profession, to be mastered by dint of application and intellectual, as well as physical, labour.

The French Revolution and the Emergence of Modern Strategies

The eruption of the French Revolution in 1789 delivered a blow to the emerging rationalistic conception of strategy from which it never quite recovered, though some of its precepts were echoed by later schools of thought, such as those of Jomini in his great work *The Art of War* (1838). The techniques of the armies of France under the Revolutionary government and later the Directory (1795–99) and Napoleon (1799–1814/15) were, superficially, those of the ancien régime—the "old order"—the political and social system of France prior to the French Revolution. Drill manuals and artillery technique drew heavily on concepts outlined in the days of Louis XVI, the last pre-Revolutionary French king. But the energy unleashed by revolutionary passion, the resources unlocked by mass conscription and a powerful state, and the fervour that followed from ideological zeal transformed strategy.

The author who understood this best was the Prussian General Carl von Clausewitz, whose military experience spanned the years from 1793 to 1815, a period in which Europe was convulsed by a series of wars centring on France. His masterpiece, *On War,* described an approach to strategy that would, with modifications, last at least through the middle of the 20th century.

As noted earlier, Clausewitz combined Enlightenment rationalism with a deep appreciation of the turbulent and uncontrollable forces unleashed by the new era. For him, strategy was always the product of tension between three poles: (1) the government, which seeks to use war rationally as an instrument of policy, (2) the military, and in particular its commanders, whose skill and abilities reflect the unquantifiable element of creativity, and (3) the people, whose animus (disposition) and determination are only partly subject to the control of the state. Thus, strategy is at once a matter of calculation and of instinct, a product of deliberation and purpose on the one hand and of emotion, uncertainty, and interaction on the other.

The wars of the mid-19th century, in particular the wars of German unification (Prussia's wars with Denmark, Austria, and France in 1864, 1866, and 1870–71, respectively) and the American Civil War (1861–65), marked a peak of Clausewitzian strategy. German Chancellor Otto von Bismarck and U.S. Pres. Abraham Lincoln successfully waged war for great stakes. Exemplary Clausewitzian leaders, they used the new instruments of the time—the mass army, now sustained year-round by early industrial economies that could ship vast quantities of matériel (equipment, apparatus, and supplies) to distant fronts—to achieve their purposes.

Yet, even in their successes, changes were already beginning to threaten the continued use of Clausewitzian strategy. Mass mobilization produced two effects: a level of societal engagement that made moderation and compromise in peacemaking difficult and conscripted armies that were becoming difficult to handle in the field. "Very large assemblies of units are in and of themselves a catastrophe," declared Prussian Gen. Helmuth von Moltke in his "Instructions for Large Unit Commanders" (1869).

Lincoln's Anaconda Plan

In 1861, as North and South gathered their forces for what would become the American Civil War, both sides prepared a grand strategy for victory. Confederate Pres. Jefferson Davis persistently adhered to a defensive strategy, permitting only occasional "spoiling" forays into Northern territory (though perhaps the Confederates' best chance of winning would have been an early grand offensive into the Union states before the North could find its ablest generals and bring its preponderant resources to bear against the South). U.S. Pres. Abraham Lincoln, on the other hand, in order to crush the rebellion and reestablish the authority of the Federal government, had to direct his blue-clad armies to invade, capture, and hold most of the vital areas of the Confederacy. His grand strategy was based on Gen. Winfield Scott's so-called Anaconda plan, a design that called for a Union blockade of the Confederacy's littoral as well as a decisive thrust down the Mississippi River and an ensuing strangulation of the South by Federal land and naval forces.

An 1861 cartoon map illustrating the Anaconda plan. Library of Congress Geography and Map Division Washington, D.C. (Digital File Number: g3701s cw0011000)

While it was the Federal armies that actually stamped out Confederate resistance, the Federal naval effort also contributed greatly to the effort. When hostilities opened, the U.S. Navy numbered 90 warships, of which only 42 were in commission, and many of these were on foreign station. By the time of Lee's surrender in 1865, Lincoln's navy numbered 626 warships, of which 65 were ironclads. From a tiny force of nearly 9,000 seamen in 1861, the Union navy increased by war's end to about 59,000 sailors, whereas naval appropriations per year leaped from approximately $12 million to perhaps $123 million. The securing of some 3,500 miles (5,600 kilometers) of coastline against daring Confederate raiders and blockade runners was a factor of incalculable value in the final defeat of the Davis government. In the last months of the war, only Galveston, Tex., remained open to the Confederates. "Uncle Sam's web feet," as Lincoln termed the Union navy, played a decisive role in helping to strangle the Confederacy.

Far from the coasts, Western waterways were major arteries of communication and commerce for the South, as well as a vital link to the Confederate states of Louisiana and Texas. Early in the war, Union strategists settled on the Mississippi and tributary rivers such as the Cumberland and Tennessee as proper avenues of attack. First to fall, in February 1862, were Fort Henry on the Tennessee and Fort Donelson on the Cumberland. After these forts fell to Union troops, Nashville was evacuated, and the way to Atlanta was clearer for Federal troops later in the war. The Battle of Shiloh, fought on April 6 and 7, 1862, farther down the Tennessee River from Fort Henry, pitted more than 100,000 men in armed struggle; after the battle, each side counted more than 1,700 dead and 8,000 wounded. The Battle of Shiloh preserved an important Union flank along the Mississippi River and opened the way to split the Confederacy along the river. In May and June of 1863, Union Gen. Ulysses S. Grant marched on Vicksburg and trapped a Southern army there. After a brilliant joint operation using land and naval forces, Vicksburg fell on July 4. With the capture of the city, the Union not only gained control of the lower Mississippi, its outlet to the Gulf of Mexico, but also effectively cut the South in two.

The Anaconda plan thus prevailed, though it was to take four years of grim, unrelenting warfare and enormous casualties and devastation before the Confederates could be defeated and the Union preserved.

Furthermore, as military organizations became more sophisticated and detached from society, tension between political leaders and senior commanders grew. The advent of the telegraph compounded this latter development; a prime minister or president could now communicate swiftly with his generals, and newspaper correspondents could no less quickly with their home offices. Public opinion was more directly engaged in warfare than ever before, and generals found themselves making decisions with half a mind to the press coverage that was being read by an expanding audience of literate citizens. And, of course, politicians paid no less heed to a public that was intensely engaged in political debates. These developments portended a challenge for strategy. War had never quite been the lancet in the hands of a diplomatic surgeon; it was now, however, more like a great bludgeon, wielded with the greatest difficulty by statesmen who found others plucking at their grip.

Compounding these challenges was the advent of technology as an important and distinct element in war. The 18th century had experienced great stability in the tools of war, on both land and sea. The musket of the early 18th century did not differ materially from the firearm carried into battle by one of the duke of Wellington's British soldiers 100 years later as they marched off to fight Napoleon; similarly, British Adm. Horatio Nelson's aptly named flagship HMS *Victory,* which saw the defeat of Napoleon's forces at the Battle of Trafalgar (1805), had decades of service behind it before that great contest. But by the mid-19th century this had changed. On land the advent of the rifle—modified and improved by the development of breech loading, metal cartridges, and later smokeless gunpowder—was accompanied as well by advances in artillery and even early types of machine guns. At sea changes were even more dramatic: steam replaced sail, and iron and steel replaced wood and canvas. Obsolescence now occurred within years, not decades, and technological experts assumed new prominence.

Military organizations did not shun new technologies; they embraced them. But very few officers had the time, or perhaps the inclination, to mull over their broader implications for the conduct of war. Ironically, this becomes clearest in the works of the great naval theorist of the age, the American Alfred Thayer Mahan. His vast corpus of work on naval history and contemporary naval affairs shaped the understanding of sea power not only in his own country but in others too, including Britain and Germany. Mahan made a powerful case that a dominant naval power, through its exercise of command of the sea, can subjugate the rest. In this respect, he argued, sea power was very different from land power: there was a vast difference between first- and second-rank sea powers but little difference between such land powers. Yet, although Mahan's doctrines found favour among leaders busily constructing navies of steam-propelled ships, all of his work rested on the experience of navies driven by sail. His theory, resting as it did on the technology of a previous era, underplayed the new and unprecedented threats posed by mines, torpedoes, and submarines. There were other naval theorists, to be sure, including the Englishman Julian Corbett, who took a different approach, emphasizing the contingent nature of maritime supremacy and the value of joint operations. However, only the group of French theorists known collectively as the Jeune École ("Young School") looked on the new naval technologies as anything other than modern tools to be fit into frameworks established in bygone times.

Strategy in the Age of Total War

It was during World War I that technological forces yielded a crisis in the conduct of strategy and strategic thought. Mass mobilization and technologies that had outpaced the abilities of organizations to absorb them culminated in slaughter and deadlock on European battlefields. How was it possible to make war still serve political ends? For the most part, the contestants fell back on a grim contest of endurance, hoping that attrition—a modern term for slaughter—would simply cause the opponents' collapse. Only the British attempted large-scale maneuvers by launching campaigns in several peripheral theatres, including the Middle East, Greece, and most notably Turkey. These all failed, although the last—a naval attack and then two amphibious assaults on Turkey's Gallipoli Peninsula—had moments of promise. These campaigns reflected, at any rate, a strategic concept other than attrition: the elimination of the opposing coalition's weakest member. In the end, though, the war hinged on the main contest in the dug-in trenches of the Western Front that snaked mostly through France near Germany's border. It was there, in the fall of 1918, that the struggle was decided by the collapse of German forces after two brilliant but costly German offensives in the spring and summer of that year, followed by a remorseless set of Allied counterattacks.

The brute strategy of attrition did not mean a disregard of the advantages offered by technology. The combatants turned to every device of modern science—from radio to poison gas, machine gun to torpedo, the internal combustion engine to aviation—to improve their abilities to make war. Peace arose, nonetheless, as a result of exhaustion and collapse, not an adroit matching of means to ends. Technology tantalized soldiers with the possibility of a decisive advantage that never materialized, while the passions of fully mobilized populations precluded compromise agreements that might have rescued the bleeding countries of Europe from their suffering.

Postwar strategic thinking concerned itself primarily with improving the art of war. To be sure, some analysts concluded that war had become so ruinous that it had lost any utility as an instrument of policy. More dangerously, there were those—the former military leader of imperial Germany Erich Ludendorff foremost among them—who concluded that henceforth war would subsume politics, rather than the other way around. And all recognized that strategy in the age of total warfare would encompass the mobilization of populations in a variety of ways, to include not merely the refinement of the mass army but also the systematic exploitation of scientific expertise to improve weapons.

Still, the keynote of the period leading up to World War II was the quest for a technological remedy to the problem of deadlock. Armoured warfare had its proponents, as did aerial bombardment. Tanks and airplanes had made a tentative debut during World War I, and, had the war lasted

The Battle of the Atlantic

At the outbreak of World War II in 1939, the primary concerns of the British navy were to defend Great Britain from invasion and to retain command of the ocean trading routes, both in order to protect the passage of essential supplies of food and raw materials for Britain and to deny those same trading routes to the Axis powers (Germany, Italy, and Japan). The German navy's goals, on the other hand, were to protect Germany's coasts, to defend its sea communications and attack those of the Allies, and to support land and air operations. The competing goals led to a great contest for control of the Atlantic sea routes. For British Prime Minister Winston Churchill, the Battle of the Atlantic had to be won by the Allies, as it represented Germany's best chance to defeat the Western powers.

The first phase of the battle for the Atlantic lasted from the autumn of 1939 until the fall of France in June 1940. During this period, the Anglo-French coalition drove German merchant shipping from the Atlantic and maintained a fairly effective long-range blockade. The battle took a radically different turn following the Axis conquest of the Low Countries, the fall of France, and Italy's entry into the war on the Axis side in May–June 1940. Britain lost French naval support at the very moment when its own sea power was seriously crippled by losses incurred in the Allied retreat from Norway and also their evacuation of more than 300,000 British, French, and Belgian soldiers from Dunkirk, France, to Great Britain before they were captured by Nazi forces. The sea and air power of Italy, reinforced by German units, imperiled and eventually barred the direct route through the Mediterranean Sea to the Suez Canal, forcing British shipping to use the long alternative route around the Cape of Good Hope at the southernmost point of Africa. This cut the total cargo-carrying capacity of the British merchant marine almost in half at the very moment when German acquisition of naval and air bases on the English Channel and on the west coast of France foreshadowed more destructive attacks on shipping in northern waters.

At this critical juncture, the United States, though still technically a nonbelligerent, assumed a more active role in the battle for the Atlantic. Through the provisions of the Lend-Lease Act, the United States aided its World War II allies with war materials, such as ammunition, tanks, airplanes, and trucks, by giving the president the authority to aid any nation whose defense he believed vital to the United States and to accept repayment with any "direct or indirect benefit which the President deems satisfactory."

And so the United States turned over 50 World War I destroyers to Great Britain, which helped to make good previous naval losses. In return, the United States received 99-year leases for ship and airplane bases in Newfoundland, Bermuda, and numerous points in the Caribbean. U.S. units were also deployed in Iceland and Greenland.

Early in 1942, after the United States had become a full belligerent, the Axis opened a large-scale submarine offensive against coastal shipping in American waters. German U-boats (submarines) also operated in considerable force along the south Atlantic ship lanes to India and the Middle East. The Allied campaign to reopen the Mediterranean depended almost entirely upon seaborne supply shipped through submarine-infested waters. Allied convoys approaching the British Isles, as well as those bound for the Russian ports of Murmansk and Archangelsk, had to battle their way through savage air and undersea attacks. It was publicly estimated at the close of 1942 that Allied shipping losses, chiefly from planes and U-boats, exceeded those suffered during the worst period of 1917 during World War I.

In 1942 and early 1943 the ever-tightening Allied blockade of Axis Europe began to show perceptible progress in combating the Axis war on shipping. With more and better equipment, the convoy system was strengthened and extended. Unprecedented shipbuilding, especially in the United States, caught up and began to forge ahead of losses, though the latter still remained dangerously high. Bombing raids on Axis ports and industrial centres progressively impaired Germany's capacity to build and service submarines and aircraft. The occupation of virtually all West African ports, including the French naval bases at Casablanca, Morocco, and Dakar, Senegal, denied to Axis raiders their last possible havens in southern waters. By these and other means, the Atlantic Allies thwarted Axis efforts to halt the passage of American armies and material to Europe and North Africa, to prevent supplies reaching Britain and the Soviet Union, and to break up the blockade of Axis Europe.

The battle's decisive stage was early 1943, when the Allies gained a mastery over Germany's submarines that translated into significant reductions in shipping losses. By the time the Allies invaded Normandy, France, on D-Day in June 1944, the Battle of the Atlantic was essentially over, and the Western powers exercised control of Atlantic sea-lanes. Though German U-boats continued to operate in the Atlantic almost until the end of the war, they were ineffective against Allied convoys and were systematically sunk almost as fast as they made it out to sea.

a little longer, they certainly would have demonstrated abilities well beyond those that were shown during the war. The advocates of armoured warfare resided for the most part in Britain, which pioneered the creation of experimental armoured forces in the early 1920s. J.F.C. Fuller in particular, a brilliant but irascible major general and the architect of what would have been the British army's war plan in 1919, made a powerful case that tanks, supported by other arms, would be able to achieve breakthroughs and rapid advances unheard of throughout most of the Great War. His voice was echoed in other countries. One such prophet was a French colonel who had spent most of the war in a German prisoner-of-war camp. Charles de Gaulle's plea for a mechanized French army (*The Army of the Future*; 1934) fell on deaf ears not so much because the French army opposed tanks (it did not) but rather because he called for a small, professional, mechanized army capable of offensive action. France's military and political leaders, accustomed to an army that had a long and deeply ingrained conscript tradition, and preferring a strategic posture of defense against invasion, was uninterested.

Herein lay the difficulty of the armoured warfare advocates in the interwar period. They saw the possibilities of an instrument for which there was no obvious use or that would run against powerful norms. The British, though anxious about imperial defense, were far less worried about Germany and allowed their armoured force to wither. The United States had the industrial tools but no conceivable use for tank divisions. The Germans were, because of restrictions imposed on them after World War I, only able to experiment in secret with tanks and their accompaniments, through 1935 at any rate. Thus, until the mid-1930s, while thinking about this new instrument of warfare proceeded, actual development of substantial field (as opposed to experimental) forces languished except among a few maverick officers.

Air warfare was a different matter. Aircraft had proven invaluable during World War I for a variety of missions—reconnaissance, artillery spotting, strafing, bombing, and even transport. All major powers rushed to acquire a variety of combat aircraft and to experiment with new types. At sea the question was one of developing the right techniques and procedures as well as technology for operating aircraft carriers. On land the issue became one of the role of aerial bombardment.

In the view of some proponents of air warfare (most notably the Italian Gen. Giulio Douhet), the advent of the long-range bomber had radically changed warfare: warfare, and hence strategy, would henceforth rest on the application of force directly against civilian targets. In some respects this was a mere extension of the idea that in total war the strategic goal was to break the will of a society to resist. Previously, however, it had been thought that this came about through the intermediary of military engagements, in which armed forces clashed until the price in blood and treasure became too great for one side to bear. Henceforth, Douhet and others argued, force-on-force had become irrelevant; in the words of British Prime Minister Stanley Baldwin, "The bomber will always get through." Not everyone acted on this belief, although few openly denied it. The fear of the effects of aerial bombardment of largely undefended cities played a powerful role in shaping public and governmental attitudes to the Munich Agreement of 1938; it did not, however, prevent countries from continuing to develop conventional land and naval forces.

The new weapons and operational doctrines—that is, the combination of organization and techniques embodied in the armoured division on land or the carrier task force at sea—were tested in World War II. This conflict represented the culmination of trends in strategic thought and behaviour manifest since the early 19th century. The mobilization of populations had become not merely total but scientific: governments managed to squeeze the last ounce of effectiveness out of men and women of all ages, who endured rationing, extended workweeks, and protracted military service to an extent unimaginable even 30 years before. Those governments that were most efficient at doing so—the U.S., the British, and to some extent the Soviet—defeated those that were less relentlessly rational. It was, ironically perhaps, the United States and Britain that adopted large-scale mobilization of women in war production and auxiliary military service, while Germany and Japan flinched at such an upheaval in social roles. In some cases, older attitudes to war, most notably a Japanese warrior ethic that paid little heed to mundane matters such as logistics or field medicine, proved dysfunctional. German and Japanese strategy often emanated from wild ideological beliefs, leading to debacles when sheer will proved unequal to carefully amassed and directed resources on the other side. As a result, strategy as a rational mode of thought seemed to triumph.

The new tools of warfare worked well, though not quite as expected. Attacks on cities and economic targets proved brutally effective, but only over time. The contest between offense and defense continued, and military leaders discovered that air forces had to win a battle against opposing airmen before they could deliver crushing force against an opponent's civilian population. On land, new formations built around the tank increased the speed of warfare and delivered some extraordinary outcomes—most notably, Germany's overthrow of France in 1940 in a campaign that was decided in less than two weeks of hard fighting and completed in less than two months.

The development of machine-centred warfare had restored mobility to the battlefield; science and the arts of administration had allowed those techniques to be fully applied.

Every auxiliary science and discipline, from weather forecasting to electronics, from abstruse forms of mathematics to modern advertising, was mobilized to its fullest. At the pinnacle, the governments that won the war did so with large, highly skilled organizations that brought together soldiers and civilians and that concluded many of the war's largest decisions in international conferences supported by hundreds, indeed thousands, of support personnel. Strategic decisions—the launching of the Anglo-American invasion of Normandy on June 6, 1944, for example—emerged through carefully weighed calculations of many kinds, from soil engineering to the intricacies of coalition politics.

Strategy in the Age of Nuclear Weapons

The period from 1939 to 1945 represented the acme of the old style of war, and with it strategy as the purposeful practice of matching military might with political objectives. In its aftermath a number of challenges to this classical paradigm of war emerged, the first in the closing days of World War II. The dropping of atomic bombs on Hiroshima and Nagasaki, Japan, inaugurated a new era of war, many observers felt. Bernard Brodie, an American military historian and pioneering thinker about nuclear weapons, declared in 1946:

> *Thus far the chief purpose of our military establishment has been to win wars. From now on its chief purpose must be to avert them. It can have almost no other useful purpose.*

Total destruction of Hiroshima, Japan, following the dropping of the first atomic bomb on Aug. 6, 1945. U.S. Air Force photo

If that were indeed the case, a strategic revolution would have occurred.

In some ways, nuclear weapons merely made effective the earlier promise of air power—overwhelming violence delivered at an opponent's cities, bypassing its military forces. Nuclear weapons were different, however, in their speed, their destructiveness, and the apparent absence of countervailing measures. Furthermore, the expense and high technology of nuclear weapons suddenly created two classes of powers in the world: those who wielded these new tools of war and those who did not.

In the ensuing decades, nuclear facts and nuclear strategy had a peculiarly uneasy coexistence. Many of the realities of nuclear weapons—how many were in each arsenal, the precise means for their delivery, the reliability of the devices themselves and of the planes, missiles, and crews that had to deliver them—were obscure. So too were the plans for their use, although a combination of declassification of early U.S. war plans and the flood of information that came out of the Warsaw Pact countries (and communist states under Soviet domination) following the collapse of the Soviet Union in 1991 illuminated some of the darkness.

Nuclear strategic thought, however, was far less murky. Those who developed it stemmed less from the military community (with a few exceptions, such as French Gen. Pierre Gallois) than from the civilian academic world and less from the discipline of history than from economics or political science. An elaborate set of doctrines developed to explain how nuclear strategy worked. One such doctrine was "mutual assured destruction" (MAD), the notion that the purpose of nuclear strategy was to create a stable world in which two opponents would realize that neither could hope to attack the other successfully and that in any war both would suffer effective obliteration.

In all cases, the centre of gravity lay with the problem of deterrence, the prevention of adverse enemy behaviours rather than concrete measures to block, reverse, or punish them. Strategic thought now entered a wilderness of mirrors: What behaviour could be deterred, and what could not? How did one know when deterrence had worked? Was it bad to defend one's population in any way—with civil defense or active defenses such as antiballistic missiles—because that might weaken mutual deterrence? The problem became more grave as additional countries acquired nuclear weapons: Were Chinese leaders deterred by the same implicit threats that worked on U.S. and Soviet leaders? For that matter, did Indians and Pakistanis view each other in the same way that Americans and Soviets viewed each other?

It is likely (although in the nature of things, unprovable) that the looming presence of nuclear weapons prevented a U.S.-Soviet conflict during the Cold War, an open yet restricted rivalry and hostility that developed after World War II between the U.S. and the Soviet Union and their respective allies.

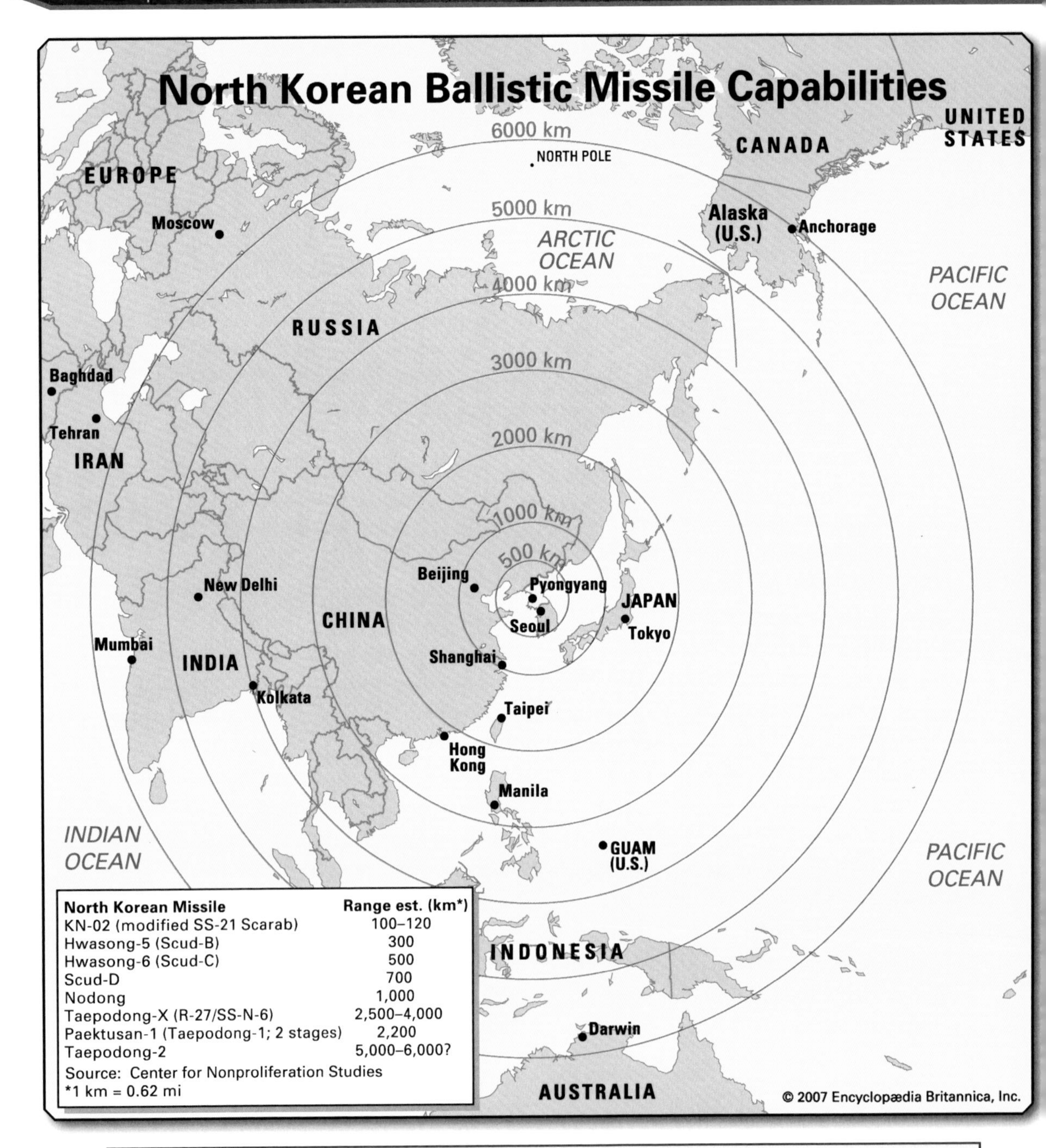

North Korean Missile	Range est. (km*)
KN-02 (modified SS-21 Scarab)	100–120
Hwasong-5 (Scud-B)	300
Hwasong-6 (Scud-C)	500
Scud-D	700
Nodong	1,000
Taepodong-X (R-27/SS-N-6)	2,500–4,000
Paektusan-1 (Taepodong-1; 2 stages)	2,200
Taepodong-2	5,000–6,000?

Source: Center for Nonproliferation Studies
*1 km = 0.62 mi

Map showing the range of North Korean ballistic missiles on an azimuthal equidistant projection centred on P'yŏngyang, as of 2007.

On the other hand, the highly probable possession of nuclear weapons by Israel in 1973 did not deter an Egyptian-Syrian conventional assault on that country. For that matter, North Vietnam seems to have disregarded American nuclear weapons during the Vietnam War (1954–75).

Initially, nuclear strategy concerned only a handful of states: the United States, the Soviet Union, China, the United Kingdom, and France. These were countries embedded, initially at least, in Cold War alliances. In 1974 India tested a nuclear device; this was followed by competitive testing of weapons with Pakistan in 1998. Israel was understood to have acquired nuclear weapons during the 1970s if not earlier, and North Korea avowed its acquisition of at least one or two weapons in 2002. In 1991 it became apparent that Iraq had a vigorous and potentially successful nuclear program, and a similar Iranian program had been under way. The spread of nuclear weapons amounted effectively to a second nuclear revolution, which may have operated by a different logic than the first. The stylized (though nonetheless frightening) standoff of the Cold War was replaced by a world in which many of the same elaborate safeguards might no longer exist, by nuclear possession on the part of countries that routinely fought one another (particularly in the Asian subcontinent), and by the development of weapons small enough to be smuggled into a country in a variety of ways. By the beginning of the 21st century then, nuclear issues had revived as a subject of strategic concern, if not serious strategic thought. The proliferation of nuclear technology by a Pakistani scientist, Abdul Qadeer Khan, and the development of nuclear weapons by Kim Jong Il's North Korea shook optimistic assumptions about the ability of the interstate system to stop marginal actors from acquiring and spreading the wherewithal to make nuclear weapons—including the possibility of terrorist groups acquiring such weapons. The overt entry of India and Pakistan into the nuclear club, the generally acknowledged Israeli nuclear arsenal, and the looming Iranian nuclear threat were no less unsettling.

Arms Control

Not surprisingly, in view of the threat of nuclear devastation, the second challenge to the traditional paradigm of strategy came from the effort to control nuclear weapons. Arms control has had a long history, perhaps as old as organized warfare itself, but it became a major feature of international politics in the interval between the two World Wars and even more so during the Cold War. A variety of agreements—from the Washington Naval Conference (1921–22) to the Anti-Ballistic Missile Treaty (1972)—constrained military hardware and forces in a variety of ways.

The theory of arms control, articulated primarily by academics, repudiated much of the logic of strategy. Traditionally, arms control has had three purposes: reducing the risks of war, preparing for the burdens of war, and controlling damages should it break out. Underlying arms control, however, lay a deeper belief that weapons in and of themselves increase the probability of armed conflict. Where Clausewitz had believed that the logic of war lay outside the realm of the forces used to wage it, arms control rests implicitly on the idea that weapons and the organizations built around them can themselves lead to conflict. Instead of war having its origins chiefly in the political intercourse of states, arms control advocates believe that war has an autonomous logic, though one that can be broken or interrupted by international agreements.

The first nuclear era, from the late 1940s through the 1990s, which was dominated by the nuclear standoff between the Soviet Union and the United States, seemed propitious for this view of the world. This was particularly true in the last quarter of the 20th century, when arms control agreements became the dominant feature of U.S.-Soviet relations and a general measure, in many parts of the world, of the prospects for peace.

The end of the Cold War meant the weakening or irrelevance of some arms control agreements, such as those that limited the distribution of conventional forces in Europe. Others were abrogated or ignored by their signatories—most notably, when the United States invoked a clause in the Anti-Ballistic Missile Treaty on Dec. 13, 2001, to withdraw from the agreement. Other conventions remained intact, though, and seemed in some cases to assume added urgency. In particular, efforts to ban chemical and biological weapons assumed new vigour, although it was not clear whether advances in science at the beginning of the 21st century would make it impossible to restrict the development of lethal toxins or artificial plagues.

The 1968 Treaty on the Non-proliferation of Nuclear Weapons (NPT) had a mixed record in blocking states from acquiring atomic or thermonuclear weapons. The NPT, coupled with energetic diplomacy by the United States and other great powers, prompted a wide range of governments, including Argentina, Australia, Sweden, and Taiwan, to terminate or put into dormancy nuclear programs. On the other hand, at least one NPT signatory, Iraq, blatantly violated the treaty with an extremely active nuclear weapons program, which was thwarted in 1981 by an Israeli preemptive attack on the nuclear reactor under construction at Osirak and thwarted again, at least for a time, by an intrusive system of United Nations inspections following the Persian Gulf War (1991).

Still, other countries have joined the nuclear club. The open acquisition of nuclear weapons by India and Pakistan (neither of which had ratified the treaty) did not diminish the prestige or importance of those countries—quite the reverse in some ways. A determined effort by North Korea to acquire nuclear weapons, even at the expense of its previous agreements with other powers, suggested that the notion of preventing proliferation by treaty or international consensus had weakened. When, in 2002, the United States formally announced a willingness to employ force preemptively against threats to its national security, more than one observer supposed this had something to do with nuclear proliferation.

The arms control critique of strategy has its greatest force in the nuclear realm because nuclear weapons are different. Even so, the logic of Clausewitzian strategy survives. Offense exists, of course, but so too does defense, in the form of anti-ballistic missiles, preemptive attack, and various forms of civil defense. States acquire weapons of mass destruction for reasons that are largely political in nature. Furthermore, international agreements remain at the mercy of states' willingness to subject themselves to them. Below a certain threshold of violence, moreover, traditional strategy still operates, as in the sparring between India and Pakistan over Kashmir. India and Pakistan, however, are both states with well-developed institutions, which means that each would have much to lose if they attempted to annihilate each other. Whether a nonstate actor, such as Hezbollah (a militia and political party in Lebanon) or al-Qaeda, would be subject to the same restraint is more questionable.

Strategy and Wars of National Liberation

In the years following World War II, scores of new states arose, mostly in Africa and Asia, many of them following protracted struggles of national liberation from European powers attempting to maintain their colonial positions. In so doing, a variety of movements and countries waged war against the technologically superior armed forces of the West. These new countries won their independence not by the force-on-force clash of conventional armies and advanced weaponry but through more subtle techniques of subversion, hit-and-run, and, often, use of terrorism.

To be sure, the European powers had faced able opponents in the past, from the indigenous Native Americans in the American colonies to the Boer farmers (descendants of Dutch farmers who had settled in what is now South Africa in the late 1600s) who battled for control of the northern part of the land with the British at the end of the 19th century. And in many ways the pattern for a successful anti-imperial force appeared shortly after the end of World War I in the form of the Irish Republican Army, which had an important role in convincing the British to end its rule in most of that island. By and large, though, the story of imperial warfare in the period before World War II was of protracted struggle leading to pacification and quiescence. Emilio Aguinaldo succumbed to the American forces in the Philippines in 1901 following a two-year rebellion; the caliph 'Abd Allāh (successor to al-Mahdī) was swept away by British rifles and machine guns in the Sudan in 1898.

Things changed dramatically after World War II. The Dutch gave up the Netherlands East Indies to a Javanese-led anticolonial movement, and Indonesia was born in 1945. Zionist rebels made

Palestine too much of a burden for British forces there, and the State of Israel was established in 1948. France yielded Vietnam to Ho Chi Minh's communists in the 1950s, and, even more painfully, gave up Algeria to the indigenous National Liberation Front in the 1960s. Portugal eventually withdrew from the mineral-rich provinces of Angola and Mozambique, which became independent nations in 1975. Even the United States was stymied by poorly equipped communist forces in Vietnam.

What changed after World War II? In some measure the transformation had occurred in the mind before being felt on the battlefield. The great powers had suffered catastrophic humiliations in Europe and, more importantly, in Asia during the war; they had lost self-confidence, and their colonial subjects had lost their sense of awe and resignation. In Europe and the United States the legitimacy of overseas rule had suffered a blow from which it could not recover: empire was no longer part of the natural order of things. At the same time, the antiliberal ideologies of Marxism-Leninism and, to a lesser extent, fascism (which lived in odd corners of the postcolonial world, primarily in Latin America, the Middle East,and South Africa) conveyed a long-term optimism about the direction history would take. There was no uniform ideology of national liberation. There was, however, a climate of opinion that pointed in the direction of new states emerging from the wreckage of the European empires, clinging with fierce pride to the emblems of independence, from airlines to general staffs, and determined to create strong centralized states that could mobilize hitherto politically inert peoples.

There was also the matter of technique and sponsorship. The greatest exponent of the new form of guerrilla warfare was the Chinese political leader and strategist Mao Zedong, who, starting in the early 1920s, drew on ancient Chinese practice as well as his own modified form of Marxism-Leninism to articulate a new strategy of revolutionary warfare. This congeries (collection) of ideas included careful grassroots political work, patience, guerrilla techniques gradually leading to conventional operations as the opposition weakened, and the selective use of terror. Others would supplement or modify Mao's thinking, but the basic concepts were given their due by Western military theoreticians, such as Roger Trinquier and Jules Roy of France, who studied revolutionary war from the other side in the 1950s.

Behind the march of revolutionary warriors, however, lay more traditional forms of military power. The Algerian insurgents against the colonial rule of France had the support of Egypt and other Arab states; the Vietnamese turned to the Chinese and Soviets for support against the United States; and the anticommunist Muslim guerrillas in the 1980s Afghan War gladly took aid from the United States to fight against the Soviet Union. State sponsorship of such movements, relatively rare in the 19th and early 20th centuries, became far more common, although impressive results (in particular the Indonesian struggle with the Dutch) also came in cases ignored by the great powers.

By the end of the 20th century, though, the post–World War II revolutionary techniques no longer appeared quite as effective as they once had. Communism had collapsed; the dogmas of Marxism-Leninism proved economically impractical, though they had at least promised ultimate victory, and confidence is a precious commodity in a revolutionary struggle. The Kurdish conflicts with Turkey and Iran yielded nothing but misery for the populations of that part of the world; only after the military power of Iraq had been shattered in 1991 was anything remotely resembling autonomy achieved for Kurds in a corner of Iraq. Palestinian guerrillas attacked Israel with increasing ferocity for decades, and again, although they inflicted suffering, it is hard to see that they achieved much that longer-term forces—demographic growth and the Israeli desire for normal state relations with its neighbours—did not. Despite tremendous efforts on both sides, vicious insurgent wars in Central America from the 1970s to the 1990s failed to overthrow a leftist regime in Nicaragua or rightist regimes in El Salvador and Guatemala. Many of the supposed advantages of the guerrilla were neutralized by skilled and even brutal opposition, external support, and above all the tenacity of classes, governments, and peoples that had no place to go.

Revolutions in the post–World War II era started as a rural phenomenon, although, as in Algiers, Algeria, in 1957, it sometimes included particularly vicious bouts of armed struggle in cities. At the end of the 20th century guerrilla warfare became more of an urban phenomenon. In countries as different as Uruguay, Algeria, Peru, and Israel, guerrilla war shifted (in many cases, it had not far to go) into pure terror directed against civilian populations. Yet here too the results often disappointed those hoping to overthrow a government or displace a population. Hiding in an apartment block differs greatly from hiding in a jungle or a wooded mountain: nature's creatures do not spy for, collaborate with, or confess to the forces of order, but human beings do.

Thus, revolutionary war proved an exceedingly powerful—and yet limited—tool. It left, however, a legacy not only in terms of geopolitics—a multiplicity of new states—but also of aesthetics and morality. The guerrilla fighter—clutching a Soviet-designed AK-47 assault rifle—was a stock figure of leftist politics in the second half of the 20th century. The legacy of terror and brutality, of violence directed against civilians as much as and often a great deal more than at soldiers, had the effect of undermining the rules by which the old strategic game had been played. Classical strategy resembled a game of chess in this respect: the pieces might have different weights and potential, but there were rules, breached occasionally but still observed, if only for the sake of convenience. As with the advent of nuclear weapons, the appearance of revolutionary war did not displace old-style militaries—countries, particularly the superpowers, still had vast arrays of tanks, submarines, jet fighters, and rocket launchers—but it raised large questions about their relevance.

Strategy and Terrorism

Revolutionary warfare often uses terror for its purposes, but terrorism has its own logic, often quite different from that of national or political groups seeking to control a state. Politically motivated terrorism, defined as the use of violence against noncombatants for the purpose of demoralization and intimidation, is an extremely old phenomenon. However, the September 11 attacks on the United States in 2001 took terrorism to a new level and opened up the possibility of a different form of warfare than any known thus far. The al-Qaeda organization that launched the simultaneous attacks on New York City and Washington, D.C., which cost some 3,000 lives and inflicted tens of billions of dollars of damage to buildings and a larger economy (particular aviation), was no traditional terrorist organization. It had its home in many countries, particularly Taliban-run Afghanistan, but it was a nonstate organization. It had its senior echelon of leaders, but these could be replaced, and it operated chiefly through terror cells proliferated around the world that could reconstitute and reshape themselves. Its aspirations, as portrayed in Osama bin Laden's declaration in February 1998 of "jihad against Jews and Crusaders," were vast and religiopolitical in nature. Al-Qaeda was, moreover, a truly global organization whose members traveled easily in a cosmopolitan world in which no place on the planet was much more than 36 or 48 hours traveling time from any other. They communicated with one another using the Internet and cellular telephones, and they reacted to international developments as portrayed on the mushrooming 24-hour-a-day, seven-day-a-week television and other news media of the new century.

Here was a final challenge to strategy as traditionally understood. The actors were no longer states but a religious movement—drawing, to be sure, not on the mainstream of Islam but a variant of it. Their final objectives (the expulsion of the United States from the Middle East and Persian Gulf and mass conversion to Islam) were on a scale well beyond any seen since World War II. And, most important, they had apparently discovered a way of bypassing the military forces of the greatest power on Earth in order to strike a more devastating blow at the American homeland than any suffered since the American Civil War of the 19th century. Furthermore, al-Qaeda had its roots in the troubles of a broader Arab, and to some extent Muslim, world that was at odds with a Western (and above all American-dominated) global socioeconomic order. Here was not a conflict among states but the spectre, if not the reality, of what American political scientist Samuel P. Huntington had called "a clash of civilizations."

For such a war, the traditional language and tools of strategy seemed radically unsuited. Indeed, the very use of the terminology of crime and punishment—"bringing the perpetrators to justice" was a common phrase—seemed to suggest that this was not war. A serious case could be made that terrorism—whether of the al-Qaeda type or any other variety—should not be regarded as war at all. Proponents of this view noted that terrorists were not organized or identified as soldiers and that they attacked civilian, not military, targets; in the case of al-Qaeda in particular, they did not even represent a state or an aspiring state. On the other hand, those who conceived of these attacks as acts of war, rather than simply as criminal acts, pointed out that they were not used for the purpose of financial gain or pure sadism but rather to achieve recognizable (if extraordinarily ambitious) political goals. The theoretical debate has not been resolved.

As a practical matter, however, the political debate was, at least for a time, resolved in the United States. The American president, George W. Bush, declared to a grieving country shortly after the September 11 attacks that the United States was indeed at war. Military power proved relevant when, within three months of the attacks, a combination of American and allied air power, special operations forces, and local allies had swept al-Qaeda's Afghan hosts out of power. In the ensuing months, although the enemy proved elusive (as guerrillas always have), a combination of covert operations, precision weaponry, and massively integrated intelligence activities enabled the United States and its allies to track down and capture or kill many key al-Qaeda operatives. This success presented another set of issues, which again remain unresolved. Should captured terrorists be treated as criminals (with rights to due process) or as prisoners of war (with a different set of rights), or should they be in some separate category? What sort of respect should countries show for one another's sovereignty in pursuing such individuals? Again, none of these problems were new; but they became particularly acute after 2001.

The problem of al-Qaeda in the early 21st century was different in other ways. Al-Qaeda was not a national movement (although it tapped ethnic and nationalist sentiments in places as different as Chechnya and Bosnia). More like a franchise, al-Qaeda was sometimes simply a source of inspiration to self-organizing groups of individuals across the globe who were united by some common beliefs and informed about technique and approach through the Internet.

In some ways one can see the rise of al-Qaeda and catastrophic terrorism (though the use of biological, chemical, and radiological weapons of mass destruction had barely begun) as a reaction to a larger development: a massive shift of the global balance of power in favour of the United States. In a series of short, sharp conflicts in the last decade of the 20th century, the United States proved to have developed armed forces in advance of any others on the planet. This reflected many advantages: a huge edge in military expenditure (nearly matching that of the rest of the world

B-2 Spirit stealth jet bomber. Northrop Grumman served as the prime contractor for the four-engine, subsonic, flying-wing aircraft, which entered operational service with the U.S. Air Force in 1993. U.S. Air Force; photo, Master Sgt. Kevin J. Gruenwald

combined), the most advanced technology in the world, a quiet revolution in training methods, and behind it all, the largest, most dynamic economy the world had ever known, accounting for somewhere between a quarter and a third of the world's production. Even great powers such as China could only hope to match the United States in a few narrowly defined areas or seek to nullify its advantages by so-called "asymmetric" means (such as guerrilla warfare).

The classical paradigm of strategy rested on a world of homogeneous forces. In Clausewitz's day, one European army looked pretty much like another; the same was true of navies as well. The one might be smaller or less efficient or slightly worse off than the other, but they used the same weapons, fought in the same formations, and thought in the same way. This basic truth held pretty much through World War II and even in large measure through the Cold War. By the 21st century, though, the vastly superior capabilities of the U.S. military had become a matter of quiet anxiety among even the general staffs of its staunchest allies.

As military power evolved through the 20th century, moreover, it became more difficult to assess. No country other than the United States, for example, could build and use a stealth intercontinental bomber. On the other hand, commercial imaging satellites at the end of the 20th century offered most governments, and even private groups, the same kind of fine-grained photographs of surface infrastructures that was once reserved for only a handful of countries. In addition, civilian communications and computing technology took the lead away from the military sector, making it difficult to measure the extent to which any country could exploit those technologies by networking computers and sensors for military purposes. Military power had become more opaque, more prone to surprises even on the part of well-credentialed analysts. To use Clausewitzian jargon, the "grammar" of war—the way in which militaries fought, the tools they could use, and the means by which they organized themselves—had changed.

So too had the logic of war. The great ideological struggles of the 20th century had ended: secular belief systems (most notably fascism and communism) had been overwhelmed or come sputtering to irrelevance. Although the idea of using military power to grab desirable pieces of territory or national resources had not ended—how else to explain Saddam Hussein's invasion of oil-rich Kuwait in 1990?—war did not seem a particularly attractive economic proposition. National prestige and honour still provided a motive for war, such as Argentina's attempt to seize the Falkland Islands, located off the coast of South America, from Great Britain in 1981), but these were isolated cases. Ethnic or religious hatred, however, persisted, as did the chaos attendant upon the collapse of states that proved incapable of maintaining themselves in the face of divisive pressures from below and corruption or gross incompetence from above.

After a brief but sincere burst of optimism following the end of the Cold War in the 1990s, subsequent experience seemed to indicate that war had changed but not vanished. Conflicts now seemed likely to take place between very different kinds of actors, and even when states confronted one another, they would use weapons unheard of in the classical period of strategy. The goals too would vary greatly, from the mundanely acquisitive to the eschatological (relating to the end of the world). Distinctions between combatant and noncombatant blurred, and even local contests would now take place before a global audience. It was all very different from anything Clausewitz had imagined.

CHAPTER 2

TACTICS

Tactics is the art and science of fighting battles on land, on sea, and in the air. It is concerned with the approach to combat; the disposition of troops and other personalities; the use made of various arms, ships, or aircraft; and the execution of movements for attack or defense. This chapter discusses the tactics of land warfare.

The word *tactics* originates in the Greek *taxis*, meaning order, arrangement, or disposition—including the kind of disposition in which armed formations used to enter and fight battles. From this, the Greek historian Xenophon derived the term *tactica*, the art of drawing up soldiers in array. Likewise, the *Tactica*, an early 10th-century handbook said to have been written under the supervision of the Byzantine emperor Leo VI the Wise, dealt with formations as well as weapons and the ways of fighting with them.

The term *tactics* fell into disuse during the European Middle Ages. It reappeared only toward the end of the 17th century, when "Tacticks" was used by the English encyclopaedist John Harris to mean "the Art of Disposing any Number of Men into a proposed form of Battle." Further development took place toward the end of the 18th century. Until then, authors had considered fighting to be almost the sum total of war; now, however, it began to be regarded as merely one part of war. The art of fighting itself continued to carry the name *tactics,* whereas that of making the fight take place under the most favourable circumstances, as well as utilizing it after it had taken place, was given a new name: *strategy*.

Since then, the terms *tactics* and *strategy* have usually marched together, but over time each has acquired both a prescriptive and a descriptive meaning. There have also been attempts to distinguish between minor tactics, the art of fighting individuals or small units, and grand tactics, a term coined about 1780 by the French military author Jacques Antoine Hippolyte, comte de Guibert to describe the conduct of major battles. However, this distinction seems to have been lost recently, and the concept of grand tactics has been replaced by the concept of the so-called operational level of war. This may be because, as will be discussed, battle in the classical sense—that is, of a pitched encounter between the belligerents' main forces—is rare in the modern era.

FUNDAMENTALS OF TACTICS

The tactics adopted by each separate army on each separate occasion depend on such circumstances as terrain, weather, organization, weaponry, and the enemy in addition to the purpose at hand. Nevertheless, while circumstances and purposes vary, the fundamental principles of tactics, like those of strategy, are eternal. At bottom they derive from the fact that, in war, two forces, each of which is free to exercise its independent will, meet in an attempt to destroy each other while at the same time attempting to avoid being destroyed.

VICTORY THROUGH FORCE AND GUILE

To achieve the double aim of destroying and avoiding destruction, opposing forces can rely on either force or guile. Assuming the two sides to be approximately equal—in other words, that neither is so strong as to be able to ride roughshod over the other (in which case tactics are hardly required)—a combination of both force and guile is necessary.

To employ force, it is necessary to concentrate in time and place. To employ guile, it is necessary to disperse, hide, and feint. Force is best generated by taking the shortest route toward the objective and focusing all available resources on one and the same action, whereas guile implies dispersion, the use of circuitous paths, and never doing the same thing twice. These two factors, most conducive to victory in battle, are not complementary; on the contrary, they can normally be employed only at each other's expense. In this way tactics (as well as strategy) are subject to a peculiar logic—one similar to that of competitive games such as football or chess but radically different from that governing technological activities such as construction or engineering, where there is no living, thinking opponent capable of reacting to one's moves. To describe this kind of logic, the American military writer Edward Luttwak has used the term *paradoxical*. The title is apt, but the idea is as old as warfare itself.

The single most effective means available to the tactician consists of putting his enemy on the horns of a dilemma—deliberately creating a situation in which he is "damned if he does and damned if he does not." For example, commanders have always attempted to outflank or encircle the enemy, thus dividing his forces and compelling him to face in two directions at once. Another example, well known to the early modern age, consisted of confronting the enemy with coordinated attacks by cavalry and cannon—the former to force him to close ranks, the latter to compel him to open them. A good 20th-century example was the World War I practice—revived by the Iraqis in their war against Iran in the 1980s—of shelling the enemy's front with a combination of high explosive and gas. The former was designed to compel him to seek cover, the latter, being heavier than air, to abandon it on pain of suffocation.

The Need for Flexibility

Thus considered, the principles of tactics look simple enough. However, it is one thing to analyze tactics in the abstract but entirely another thing to put theory into practice under different circumstances, on different kinds of terrain, against different kinds of enemy, with the aid of troops who may be tired or confused or recalcitrant, and amid every kind of mortal danger. As the great Prussian general Carl von Clausewitz said, "In war everything is simple, but even the simplest thing is difficult." Sophisticated tactics require well-trained, articulated forces consisting of different units that are armed with different weapons and possess different capabilities. Next, these units must be subordinated to a single directing brain and must be employed in a coordinated manner following a single, well-considered plan: hence the principle of unity of command.

Even then, tactics are not just a question of executing a plan, however clever and well conceived. In tactics, even more than elsewhere, a commander who can only make a plan and carry it out avails nothing; inasmuch as he is confronting a living enemy, what matters is his ability to adapt the plan to that enemy's reactions rapidly, smoothly, and without losing his grip. Flexibility is thus a cardinal principle of tactics. But flexibility will prevail only if it can be bound by a firm disciplinary framework. Moreover, flexibility and discipline are not easy to combine and can often be achieved only at each other's expense. Other things being equal, the larger and more powerful a given force, the less flexible it will be.

As an armed force exchanges blows with an enemy, adapting to his moves and forcing him to adapt in return, opportunities to take him by surprise should present themselves. Surprise presupposes secrecy, but secrecy may be hard to combine with the rapid action that is often necessary for implementing surprise. Like everything else in tactics, overcoming this paradox is a matter of striking a balance, first in general and then against a specific enemy, under specific circumstances and with a specific objective in mind.

The Importance of Terrain

Finally, in tactics (as in strategy) there is the topographical element to consider.

Land warfare is fought neither in a vacuum nor on a uniformly checkered board. Instead, it unfolds over concrete terrain, including roads, passages, elevated ground, cover, and obstacles of every kind. Victory goes to him who best understands and utilizes the terrain; this may be done by, for example, occupying dominant ground, utilizing cover, compelling the enemy to fight on terrain for which his forces are not suitable, cornering him, outflanking him, or surrounding him. All these methods are as old as warfare, yet at the same time they remain relevant to the present age. On their correct application the outcome of battle depends.

HISTORICAL DEVELOPMENT

Though the principles of tactics are eternal, they have been applied over history to conditions that have varied from tribal disputes to confrontations between nuclear-powered states.

TRIBAL AND ANCIENT TACTICS

It cannot be said that the warfare of the first societies was "scientific," in the sense of being consciously based on principles that are applied rationally to existing conditions. But eventually, through trial and error and the practice of increasingly clever tactics, warfare became a science and an art—particularly in the era of classical Greece and Rome.

THE AMBUSH AND THE TRIAL OF STRENGTH

The oldest, most primitive field tactics are those that rely on concealment and surprise—i.e., the ambush and the raid. Such tactics, which are closely connected to those used in hunting and may indeed have originated in the latter, are well known to tribal societies all over the world. Typically the operation gets under way when warriors, having reconnoitred the terrain and stalked their enemy, take up concealed positions and wait for the signal. The engagement opens by means of such long-range missile weapons as the javelin, the bow, the sling, and the tomahawk. Once the enemy has been thrown into disorder and some of his personnel killed or wounded, cover is discarded, and short-range weapons such as club, spear, and dagger are employed for delivering the coup de grace. Since concealment is vital and there is no sophisticated logistic apparatus, the number of combatants is usually no more than a few dozen or, at the very most, a few hundred. Tactical units are unknown and command arrangements, to the extent that they exist at all, elementary. None of this, however, is to say that such tactics are simpleminded. On the contrary, making the best use of difficult terrain such as mountains, forests, or swamps usually requires much skill and presupposes an intimate familiarity with the surroundings.

Apart from ambush and raid, which depend on making the best possible use of terrain, many primitive tribes also engage in formal, one-to-one frontal encounters that are part battle, part sport. The weapons employed on such occasions usually consist of the club (or its more advanced form, the mace), spear, and javelin, sometimes joined by the bow and special blunted arrows. Defensive armour consists of nonmetallic body cover of wood, leather, or wickerwork, often made in fantastic forms and painted extravagant colours in order to enlist the aid of spirits and terrify the opponent. Such fights differ from those described above in that the warriors stand in full view of each other across specially selected level terrain, the objective being to please the spectators and gather glory for themselves. However, here too there can be no question either of formations or of a command system. Rather, each man picks his opponent and fights separately. Hence, it is impossible to speak of tactics, except in the limited sense of the skill displayed by individual warriors in handling their weapons.

The Phalanx

To judge from numerous descriptions in Homer, archaic Greek warriors still acted in this way. The heroes on each side knew each other by reputation and sought each other out, forming pairs and fighting hand-to-hand without any regard for either collective action or the discipline and organization that were needed for it. However, the *Iliad* also contains passages that may indicate a more advanced form of tactics—namely, the phalanx. Phalanx tactics are known from ancient Sumer and Egypt as well as from Greece. Their essence consisted of packing troops together in dense, massive blocks, to some extent sacrificing flexibility, mobility, and the possibility of concealment in order to achieve mutual protection and maximize striking power. In Greek armies the usual number of ranks was 8, but formations 16 and even 50 deep are recorded. Insofar as they relied on brute force, such tactics were often considered primitive even in their own day—for example, by the Persian commander Mardonius in describing them to his master, Xerxes I. For a phalanx to execute even a simple lateral (sideways) evasive move, the troops had to be "professors of war"; such was the Roman historian Plutarch's expression in describing the disaster suffered by Sparta at the Battle of Leuctra in 371 BCE. In that battle, Spartan forces were outwitted by Theban general Epaminondas, who used an unusual battle formation to concentrate his forces against the enemy's command, killing Sparta's King Cleombrotus. The rigidity of the phalanx form made it difficult for the Spartans to react quickly to the Theban attack. As Sumerian reliefs, Egyptian wooden models, and Greek narratives show, the typical weapons employed by the phalanx were consistently short-range, hand-held instruments such as sword, spear, and pike, used in accordance to whether individual duels or mass action was considered more important. These weapons were invariably combined with defensive gear such as helmets, corselets, shields, and greaves (shin armor), although here too the amount of protection varied from one case to the next.

Detail from the Great Temple of Ramses II at Abu Simbel in Egypt showing Ramses II on a chariot. G. Dagli Orti/De Agostini/Getty Images

THE CHARIOT

Invented in the 3rd millennium BCE, the first chariots seem to have been too slow and cumbersome to serve in combat, but about 2000 BCE the light, horse-drawn, two-wheeled vehicles destined to revolutionize tactics appeared in the Western Steppe and Mesopotamia, Syria, and Turkey, from which they spread in all directions. In combination with the bow, the chariot represented a very effective system, so much so that in biblical times it became almost synonymous with military power. The great advantage of the chariot was its speed, which permitted it to drive circles around the phalanx, staying out of range while raining arrows on the foot soldiers. Once the latter had been thrown into disorder, it might be possible to put the chariots into formation, charge, and ride the enemy down. The chariot's principal drawbacks were its expense and unsuitability for difficult terrain. Also, it made inefficient use of manpower, since each vehicle required a crew of two and sometimes three men—only one of whom actually handled offensive weapons and struck at the enemy.

Light and Heavy Cavalry

The next development following chariots was cavalry, which took two forms. From Mongolia to Persia and Anatolia—and, later, on the North American plains as well—nomadic peoples fought principally with missile weapons, especially the bow in its short, composite variety. Equipped with only light armour, these horsemen were unable to hold terrain or to stand on the defensive. Hence, they were forced to employ their characteristic highly mobile "swarming" tactics, riding circles around the enemy, keeping their distance from him, showering him with arrows, engaging in feigned retreats, luring him into traps and ambushes, and forming into a solid mass only at the end of the battle with the aim of delivering the coup de grace. Being obliged to keep their possessions few and light, nomads typically were unable to compete with sedentary civilizations in general material development, including not least metallurgy. Nevertheless, as the Mongols' campaigns were to show, their war-making methods, natural hardihood, and excellent horsemanship made them the equal of anyone in either Asia or Europe until at least the end of the 13th century CE.

Far earlier, among the technically more advanced sedentary civilizations on both edges of the Eurasian landmass, a different kind of cavalry seems to have emerged shortly after 1000 BCE. Reliefs from great Assyrian palaces show horsemen, clad in armour and armed with spear or lance, who were used in combination with other troops such as light and heavy infantry. The function of these cataphracts (from the Greek word for "armour") was not to engage in long-distance combat but to launch massed shock action, first against the enemy cataphracts and then, having gained the field, against the enemy foot. The fact that ancient cavalry apparently did not possess the stirrup has often led modern historians to question the mounted soldier's effectiveness. They argue that, since riders held on only by pressure of their knees, their ability to deliver shock was limited by the fear of falling off their mounts. This argument fails to note that, particularly in Hellenistic times and again in late Roman ones, cavalry forces did indeed play an important, often decisive, part in countless battles. Still, it is true that never during classical antiquity did cavalry succeed in replacing the formations of heavy infantry that remained the backbone of every army.

Combined Infantry and Cavalry

Classical Greek warfare, as mentioned above, consisted almost exclusively of frontal encounters between massive phalanxes on both sides. However, about the time of the Peloponnesian War (431–404 BCE), the phalanx became somewhat more articulated. This permitted the introduction of elementary tactical maneuvers such as massing one's forces at a selected

BATTLE OF GAUGAMELA

On Oct. 1, 331 BCE, this clash between the forces of Alexander the Great of Macedonia and Darius III of Persia decided the fate of the Persian empire. Attempting to stop Alexander's incursion into the Persian empire, Darius prepared a battleground on the Plain of Gaugamela, near Arbela (present-day Irbil in northern Iraq), and posted his troops to await Alexander's advance. Darius had the terrain of the prospective battlefield smoothed level so that his many chariots could operate with maximum effectiveness against the Macedonians. His total forces greatly outnumbered those of Alexander, whose forces amounted to about 40,000 infantry and 7,000 cavalry.

Alexander's well-trained army faced Darius's massive battle line and organized for attack, charging the left of the Persians' line with archers, javelin throwers, and cavalry, while defending against Darius's outflanking cavalry with reserve flank guards. A charge by Persian scythed chariots aimed at the centre of Alexander's forces was defeated by Macedonian lightly armed soldiers. During the combat, so much of Darius's cavalry on his left flank were drawn into the battle that they left the Persian infantry in the centre of the battle line exposed. Alexander and his personal cavalry immediately wheeled half left and penetrated this gap and then wheeled again to attack the Persians' flank and rear. At this Darius took flight, and panic spread through his entire army, which began a headlong retreat while being cut down by the pursuing Greeks. The Macedonian victory spelled the end of the Persian empire founded by Cyrus II the Great and left Alexander the master of southwest Asia.

point, outflanking the enemy, and the oblique approach (in which one wing stormed the enemy while the other was held back). In addition, the phalanx began to be combined with other kinds of troops, such as light infantry (javelin men and slingers) and cavalry. Indeed, the history of Greek warfare can be understood as a process by which various previously existing types of troops came to be combined, integrated, and made to support one another. This development gained momentum in 4th-century BCE battles, and it culminated in the hands of the great Macedonian conqueror Alexander III the Great, whose army saw most of these different troops fighting side by side. The major exception was horse archers, which were incompatible with a settled way of life and which never caught on in the West. Another was the chariot, which was already obsolescent and, except in backward Britain, disappeared almost completely after it proved to be an ineffective tool of war at the Battle of Gaugamela in 331 BCE.

Alexander died in 323 BCE , but after his death, his successors (*diodochoi*) continued his practice of commanding standing armies consisting of professionals. Both Alexander and the *diodochoi* operated on a much greater scale than did most of their predecessors. The most important *diodochoi* were quite capable of concentrating 80,000 to 100,000 men at a single spot. These armies typically went into battle with a force of light infantrymen in front (elephants were sometimes used, but on the whole they proved as dangerous to their own side as to the enemy). Behind the light troops came the heavy phalanx, flanked by cavalry on both sides. The action would start with each side's light troops trying to drive the opponents back upon their phalanx, thus throwing it into disorder. Meanwhile, the cavalry stood on both sides. Usually one wing, commanded either by the king in person or by one of his closest subordinates, would storm forward. If it succeeded in driving away the opposing cavalry—and provided it remained under control—it could then swing inward and act as the hammer to the phalanx's anvil.

The Roman Legion

Though its exact origins are unknown, the Roman legion seems to have developed from the phalanx. In fact, it was a collection of small, well-integrated, well-coordinated phalanxes arrayed in checkerboard formation and operating as a team. Hellenistic heavy infantry relied on the pike almost exclusively; the legion, by contrast, possessed both shock and firepower—the former in the form of the short sword, or *gladius*, the latter delivered by the javelin, or *pilum*, of which most (after 100 BCE, all) legionnaires carried two. Screening was provided by light troops moving in front, cohesion by pikemen in the third and rearmost rank. Short arms made it easier for individual soldiers or subunits to turn and change direction. Too, careful articulation, a well-rehearsed command system, and the use of standards—which do not seem to have been carried by Hellenistic armies—made the legion a much more flexible organization than the phalanx. No Greek army could have imitated the movement carried out by Caesar's troops at Ruspinum in Africa in 47 BCE, when part of a legion was made to turn around and face an enemy cavalry force coming from the rear. As numerous battles showed, where the terrain was uneven and the chain of command broke down, the legion's advantage was even more pronounced. A phalanx whose ranks were thrown into disorder and penetrated by the enemy's infantrymen was usually lost; a legionary commander could rely on his soldiers' swords to deal with intruders, meanwhile bringing up additional units from both flanks.

As a formation whose main power consisted of its heavy infantry, the legion remained unmatched until the introduction of firearms and beyond. Attempts to imitate its armament and methods were made right down to the 16th century, and even today some countries still call their forces legions in commemoration of its prowess. During the 1st century BCE, legionary organization underwent some changes at the hands of generals and politicians Gaius Marius and Lucius Cornelius Sulla until it reached the zenith of its development about the time of

Julius Caesar. Subunits became larger, and the legion incorporated a detachment of heavy cavalry as well as field artillery in the form of catapults—thus turning into a combined-arms unit and becoming a true forerunner of the modern division. Yet the legion, too, had its limitations when it came to fighting in the dense forests of Germany or, even more so, the open deserts of the Middle East. The Romans were eager to defeat the Parthian Empire (located in what is now Iran), but were unable to do so, especially after Roman leader Marcus Licinius Crassus's disastrous defeat at Carrhae in 53 BCE. In that battle, as Crassus crossed the desert east of the Euphrates River, suddenly the Parthians were upon him, with a force of about 1,000 armoured knights and nearly 10,000 horse archers. His troops were neither acclimatized nor adapted to desert warfare. While his son Publius in vain launched a covering attack with his cavalry, the main Roman forces had formed a square against the encircling Parthians and tried unsuccessfully to cover both body and head with their shields against the showers of Parthian arrows. The Parthians' provision of a corps of 1,000 Arabian camels, one for every 10 men, enabled the Parthians to retire by sections and replenish their quivers. Crassus, lacking provisions, was compelled by his demoralized men to negotiate but was cut down by the Parthians in the attempt. About 10,000 Romans escaped, but the rest of Crassus's men were either captured or killed. The Parthians had dealt a stunning blow to Roman prestige in the East.

The lesson was not lost. From the 6th century CE, the Byzantine army always supplemented its infantry and heavy cavalry with units of horse archers, usually consisting of mercenaries recruited from various barbarian tribes. In this way, they were able to counter the Arabs and, later, the Seljuq Turks.

Medieval Tactics in the West

Whatever their differences, Byzantine armies were the direct heirs of the Roman legions in that they consisted of various kinds of troops in well-organized, centrally commanded units. Meanwhile, developments in the Latin West followed a different course.

The Barbarians

The Germanic peoples who finally brought down the Western Roman Empire in the year 476 CE (the Eastern Roman Empire became known as the Byzantine Empire and lasted for another thousand years) were formidable foot soldiers more notable for physical prowess and courage than for tactical organization. Weapons were mostly hand-held and included the sword, spear, and javelin. To these the Franks (one of these Germanic peoples) added the heavy battle-axe, or *francisca*, useful for both hacking and throwing. Defensive arms consisted of the usual helmets, corselets, greaves, and shields—although, since metal was expensive, most warriors seem to have worn only light armour. Sources mention the names of some tactical formations such as the hogshead, which apparently consisted of phalanxlike heavy blocks, but movement may have been carried out in smaller units, or *Rotte*. Germanic formations and tactics must have been effective, for in the end they overcame—or rather superseded—the Roman legions; how it was done, though, simply is not known.

The Mounted Knight

Dominating present-day northern France, Belgium, and western Germany, the Franks established the most powerful Christian kingdom of early medieval western Europe. The name France (Francia) is derived from their name. If sources can be trusted, the Franks still fought mainly on foot when they defeated the Muslim forces invading from Spain at Poitiers in 732 CE. But about the time of the great King Charlemagne, later in the 8th century—and possibly aided by the stirrup, which was introduced to Europe from the East—they took to horse and became knights. Typically, knights carried elongated, kite-shaped shields and wore a complete suit of metal armour (sometimes the horse too was armoured). Their principal offensive weapon was the lance. Originally, this was comparatively light and short, and it could either be held overhead (or even thrown, as shown in the Bayeux Tapestry) or else gripped underhand parallel to the horse's body. However, about the year 1100 the technique of couching the lance under the arm was introduced. This permitted it to grow much longer and heavier and also meant that knights were becoming more specialized for fighting other knights. The secondary weapon was the sword, which, like the lance, tended to grow longer and heavier with time. Knights would open combat with the lance and continue it with the sword, fighting either on horseback or, if forced to dismount, on foot. In time, chain-mail armour tended to be replaced by stronger, but less flexible, plate. The new suits, which steadily grew heavier, rendered their wearers less capable of dismounted action and, as legend has it, allowed them to get on horseback only with the aid of a crane.

By virtue of their mobility, height above the ground, and sheer weight, knights possessed a tremendous advantage over foot soldiers, especially those caught on open terrain and not operating in organized formations. Though social differences among knights were very great, in principle each regarded himself as militarily the equal of every other. In addition, since feudal armies were made up entirely of officers, as it were, they tended to be ill-organized, ill-disciplined, and prone to sedition. Only on occasion were there attempts at tactical organization and a regular chain of command. If modern reconstructions can be trusted, armies might enter battle in an orderly manner, usually operating in three divisions with the commander in chief in charge of the rearmost one. However, medieval princes such as Harold II of England, William I the Conqueror, and Richard I the Lion-Heart were expected to engage in hand-to-hand combat or else, by showing cowardice, lose standing in the eyes of their subordinates. Therefore, it was seldom long before engagements ran out of control and degenerated into cavalry melees. Fighting as individuals or in small groups, knights clumped together and hacked away indiscriminately at each other. Since armour was heavy and quarter usually given (to be followed by the payment of ransom), casualties among the chivalry were often light. One side having succeeded in killing, capturing, or driving off the other's horsemen, the foot soldiers present would be slaughtered like cattle.

The European system centring on armoured shock cavalry was only moderately effective when faced with the swarming horse archers of the East. Against the Saracens (Muslims) during the Crusades, for example, it was capable of holding its own—provided the knights were kept on a tight rein and did not allow themselves to lose cohesion, become separated from the foot soldiers, or fall into an ambush. Such methods gave good results when employed by Richard the Lion-Heart in the Battle of Arsūf in 1191; in that battle against the great Muslim leader Saladin, after the crusaders had left Arsūf, the Muslim attacks became more intensive and were concentrated against the Hospitalers, Richard's rear guard. Richard forbade them to counterattack until the evening, then launched a general charge that overwhelmed Saladin's army and inflicted heavy losses on the forces attacking to the rear. Seven hundred crusaders and several thousand Muslims were killed. However, when necessary precautions were not taken and interarm cooperation broke down, the outcome could well be disastrous defeat, as at Ḥaṭṭīn four years earlier. In the 13th century CE, the Mongol empire arose in the steppes of central Asia and began an aggressive war of conquest against its neighbours, from China to the south to Russia and other eastern European lands to the west. Employed against the Mongol invaders of Europe, knightly warfare failed even more disastrously for the Poles at Legnica and the Hungarians at Mohi in 1241. But most of feudal Europe was saved from sharing the fate of China and the Russians not by its tactical prowess but by the unexpected death (in 1241) of the Mongols' supreme ruler, Ögödei, and the subsequent eastward retreat of his armies.

Nevertheless, within Europe itself for a period of perhaps three centuries, the best and indeed almost the sole means of stopping one troop of armoured cavalry was another troop of armoured cavalry.

Bowmen and Pikemen

The first field tactics that proved capable of countering the knight were built around the bow and the crossbow. Both might be used either in difficult terrain or from behind some artificial obstacle such as pits (as at Bannockburn in 1314), stakes (as at Crécy in 1346, Poitiers in 1356, and Agincourt in 1415), or a trench dug in the earth. The bow in its most powerful form, the longbow, was a cheap, low-class weapon originally associated with primitive social organizations such as the Welsh tribes. The crossbow, a much more expensive and sophisticated weapon, was typically employed by urban militias and mercenaries. The two weapons' technical characteristics were somewhat different, especially as regards the crossbow's shorter range, lower rate of fire, and greater penetrating power; as a result, they were seldom seen side by side in the same battle. However, both were capable of defeating armour, even the heavy plate worn toward the close of the Middle Ages, and were therefore useful against knights when properly employed. Proper employment meant selecting suitable positions and forming long, thin formations, sometimes in the form of a shallow W in order to trap attackers and enfilade them. Because formations such as these were difficult to move from place to place, they and the weapons on which they were based were better suited for the defense than for the offense.

This particular disadvantage was not shared by two other nonchivalrous weapons, the halberd (a weapon consisting of an ax blade balanced by a pick with an elongated pike head at the end of the staff, about 5 feet [1.5 metres] long), and pike (a long spear with a heavy wooden shaft 10 to 20 feet [3 to 6 metres] long, tipped by a small leaf-shaped steel point). They became the specialty of the Swiss, who, because of the mountainous terrain where they lived and economic reasons, never had much use for horses and knightly trappings. A *Haufe* (German: "heap") of Swiss infantry had much in common with a Macedonian phalanx, except that it was smaller and more maneuverable. Most of the troops seem to have been lightly armoured, wearing helmet and corselet but not being burdened by either greaves or shield. Hence, they possessed good mobility and formidable striking power. The first shock would be delivered by the pikes sticking out in front, after which the halberdiers would leave formation to do their deadly work. The Swiss differed from the Macedonians in that they did not combine the phalanx with cavalry but relied on infantry for both fixing the enemy and striking him. Usually they entered battle in three columns moving independently, thus permitting a variety of maneuvers as well as mutual support. An enemy could be engaged from the front, then hit in the flank by a second *Haufe* following the first in echelon formation (an arrangement of a body of troops with its units each somewhat to the left or right of the one in the rear like a series of steps).

Though it is hard to be certain, apparently the hard-marching Swiss possessed sufficient operational mobility to keep up with cavalry, at any rate in confined terrain such as Alpine valleys. If the worst occurred and an isolated column was caught in the open, the troops could always form a square or hedgehog, facing outward in all directions while keeping up a steady fire from their crossbows and relying on their pikes to keep the opposing horse at a respectful distance until help arrived. Whereas the Scots inhabited a northern wilderness, the Swiss were located in the centre of Europe, and, whereas the Flemish went down in front of French chivalry at Roosebeke in 1382, the Swiss won a series

LONGBOWMEN AT AGINCOURT

In August 1415, during the Hundred Years' War, Henry V of England, in pursuit of his claim to the French throne, assembled an army of about 11,000 men and invaded Normandy. The English took Harfleur in September, but by then half their troops had been lost to disease and battle casualties. Henry decided to move northeast to Calais, an English enclave in France, whence his diminished forces could return to England. However, large French forces under the constable Charles I d'Albret blocked his line of advance to the north.

On Oct. 25, 1415, the French force, which totaled 20,000 to 30,000 men, many of them mounted knights in heavy armour, caught the exhausted English army at Agincourt (now Azincourt in Pas-de-Calais département). The French unwisely chose a battlefield with a narrow frontage of only about 1,000 yards of open ground between two woods. In this cramped space, which made large-scale maneuvers almost impossible, the French virtually forfeited the advantage of their overwhelming numbers. At dawn, the two armies prepared for battle. Three French divisions, the first two dismounted, were drawn up one behind another. Henry had only about 5,000 archers and 900 men-at-arms, whom he arrayed in a dismounted line. The dismounted men-at-arms were arrayed in three central blocks linked by projecting wedges of archers, and additional masses of archers formed forward wings at the left and right ends of the English line.

Henry led his troops forward into bowshot range, where their long-range archery provoked the French into an assault. Several small French cavalry charges broke upon a line of pointed stakes in front of the English archers. Then the main French assault, consisting of heavily armoured, dismounted knights, advanced over the sodden ground. At the first clash the English line yielded, only to recover quickly. As more French knights entered the battle, they became so tightly bunched that some of them could barely raise their arms to strike a blow. At this decisive point, Henry ordered his lightly equipped and more mobile English archers to attack with swords and axes. The unencumbered English hacked down thousands of the French, and thousands more were taken prisoner, many of whom were killed on Henry's orders when another French attack seemed imminent.

The battle was a disaster for the French. The constable himself, 12 other members of the highest nobility, some 1,500 knights, and about 4,500 men-at-arms were killed on the French side, while the English lost less than 450 men. The English had been led brilliantly by Henry, but the incoherent tactics of the French had also contributed greatly to their defeat.

of spectacular victories at Morgarten (1315), Laupen (1339), Sempach (1386), and Granson (1476). These two factors combined to give Swiss tactics a reputation in Europe. From about 1450 to 1550, every leading prince either hired Swiss troops or set up units, such as the German *Landsknechte*, that imitated their weapons and methods—helping to bring down the entire feudal order.

Inferiority of Medieval Tactics

Compared to the most powerful ancient armies, however, even late medieval ones were impermanent and weak. Numbers never approached those fielded during Hellenistic and Roman times: it was a mighty medieval prince who could assemble 20,000 men (of whom perhaps 5,000 would be knights), and most forces were much smaller. Apart from the stirrup, an invention whose importance may have been exaggerated by modern historians, no important advances took place in military technology. Consequently, tactics tended to repeat themselves in cycles rather than undergo sustained, secular development—as was to become the case after 1500 and, above all, after 1830. If only because medieval discipline was often lax and organization usually elementary, sophisticated tactical maneuvers such as those which characterized the armies of Alexander, his Hellenistic successors, and the Romans at their best were few and far between. Otherwise put, the knightly system of making war was much more individualistic than its classical predecessors; had the two been pitted against each other, the earlier forms would likely have overcome the later.

The Advent of Firearms

Gunpowder apparently reached Europe from the East shortly before 1300, and firearms appeared during the 14th century. Throughout the 15th century firearms and crossbows continued to be used side by side. The first battles actually to be decided by firearms were fought between French and Spanish troops on Italian soil early in the 16th century; these included Marignano (1515), Bicocca (1522), and, above all, Pavia (1525).

Adaptation of Pike and Cavalry Tactics

The first firearms were primitive devices lacking both buttstock (for bracing against the shoulder) and trigger; hence, they had to be held under the arm and could scarcely be aimed. It was only during the second half of the 15th century that the harquebus, which incorporated both of these features, made its appearance. This was a great improvement, but the harquebus still suffered from a low rate of fire as well as inaccuracy and unreliability. In order to compensate for these disadvantages and build staying power, 16th-century units such as the famous Spanish tercio were made up of pikemen surrounded by "sleeves" of harquebusiers on each corner. Much like the light armed troops of antiquity and the crossbowmen who accompanied the Swiss *Haufen*, harquebusiers would open the action and then retreat behind the pikemen as the latter came to close quarters with the enemy. Hence, 16th- and early 17th-century battles still tended to be decided by "push of pike," as the saying went.

In the face of such formations, lance-carrying cavalry operating on its own was almost helpless. During the 16th century, an attempt was made to adapt cavalry to the new circumstances by arming it with short firearms such as pistols and carbines. These were difficult to load on horseback and had neither the range nor the accuracy to permit Mongol-style swarming tactics. Instead, the cavalrymen carrying them were trained to attack infantry formations by approaching them in serried ranks, firing at point-blank range, and withdrawing in turn—a maneuver resembling the orderly moves of a ballroom dance and known as the caracole. Insofar as they sacrificed the cavalry's greatest advantages—namely, its mobility and sheer mass—such methods were never very effective. A much better system was to rely on combined arms, bombarding infantry formations with artillery (another 14th-century invention that began to make its impact felt on the battlefield from about 1500) and then, once the infantry had been shattered, sending in the heavy cavalry to complete the job with cold steel.

The State-Owned Army

As European firearms improved, the old situation in which each people possessed its own weapons and, therefore, its own system of organization and tactics disappeared. From about 1600, so great was the superiority of European arms and military methods that non-European societies could survive, if at all, only by excluding or imitating them. Inside Europe, too, armies and tactics became increasingly alike. Gone were the days when one nation specialized in heavy cavalry, another in light cavalry, still another in pikemen, archers, or crossbowmen. Everywhere armed forces were becoming divorced from society at large and growing into regular, state-owned organizations that tended to resemble one another. These similarities were reinforced by the international character of warfare, which for centuries on end permitted individuals and even entire units to move from one service to another.

During the second half of the 16th century, every army came to consist of three arms: infantry, cavalry, and artillery. The trend was to add more and more of the first and third arms, while the second, though retaining its high social prestige, underwent a relative decline in numbers and importance. By the early 18th century a fourth arm, engineering, had differentiated itself from artillery, an arrangement that became standard in all armies after the Seven Years' War (1756–63). Particularly after 1683, the year in which the Turks mounted their last major challenge and were repulsed at the gates of Vienna, European armies grew accustomed to seeing one another as their strongest opponents. Since they organized, trained, and equipped themselves to fight one another, there was a tendency to distinguish *les grandes opérations de guerre* from *guerrilla*, or small war, which was increasingly left to so-called free corps, or irregulars. As the regulars came to rely on heavier and heavier weapons, the gap between the two kinds of warfare grew. Ultimately, this specialization was to cost armies the ability to fight opponents that did not resemble themselves, but in the 17th century that development still remained far in the future.

Linear Formation

Meanwhile, the improvement of firearms caused armour to be discarded. Infantry ceased wearing it almost completely after 1660, and the armour carried by cavalrymen grew steadily shorter until all that remained were the breastplates worn by heavy cavalry—the cuirassiers—as late as the 20th century. The harquebus developed into the heavier, more powerful musket, which soon acquired the flintlock firing mechanism. This was scarcely the perfect weapon, but it could be relied on to fire two or three times per minute to an effective range of 100–150 yards (91 to 137 metres) without misfiring more than 20 percent of the time. There was a constant tendency to increase the number of musketeers at the expense of pikemen until, by the end of the Thirty Years' War (1618–48), their proportions had become about equal. To allow the maximum number of barrels to fire without mutual interference, tactical units grew smaller, and the number of ranks drawn up behind one another declined. From 8 to 10 at the time of Prince Maurice of Nassau early in the 17th century, it came down to 4 or 5 a century later, 3 or 4 in the armies of Frederick the Great, and 2 or 3 toward the end of the 18th century.

To maximize efficiency, drill was invented. It first made its impact felt in the Dutch army under Maurice of Nassau, a great teacher whose headquarters attracted aspiring officers from all over Europe. Standards, often modeled after Roman ones, were introduced to help units align themselves, and tactical movements were carried out to the sound of trumpets, bugles, and drums—the latter an Oriental innovation apparently brought to Europe about 1500. In this age of René Descartes, Thomas Hobbes, and Louis XIV, each of whom in his different way was determined to reduce the world to order, the military ideal was to achieve maximum reliability and efficiency by training troops to operate in a machine-like manner. This implied much tighter discipline and organization, which in turn required a shift toward the type of regular, professional forces that alone were capable of achieving them.

About 1670 the bayonet was invented, causing pikes to be discarded and homogeneous infantry to be created (though the expression "to trail a pike" lingered for another century). Apart from predicaments when it had to form squares in order to confront attacking cavalry, infantry now fought in very long, thin formations. Throughout the 18th century a lively debate was carried on concerning the best ways to employ these formations, but basically each side organized its forces in two lines separated by perhaps 300 to 400 yards (275 to 365 metres) and moving forward one behind the other. Though the precise arrangements varied from one army to the next, inside each line the units were organized by platoon, company, and battalion. Advancing toward each other, each side would hold its fire for as long as possible in order to close range and obtain a better aim, and then, acting upon the word of command, the opposing lines would fire salvo after salvo into each other. The final step consisted of fixing bayonets and storming the enemy—although, since one side usually broke, actual hand-to-hand fights seem to have been rare. Flank protection was provided by light cavalry such as dragoons or hussars, which were introduced in force between 1690 and 1740. Heavy cavalry would be held in reserve, ready to strike when a gap was created or a flank presented itself. During the second half of the 18th century another type of cavalry, the lancers, was added specifically to root out gunners hiding under their cannons' barrels.

The first cannon were slow-firing devices much too cumbersome to take part in tactical maneuvers, and indeed so heavy were they that until about 1500 they were not even provided with wheels. Even then, the standard method was to position the guns in the intervals between units and in front of the advancing lines. This permitted them to open the battle but subsequently forced them to fall silent as the army advanced and left the gunners behind. To solve this problem, there was a steady tendency to make artillery smaller and more mobile, from the "leather guns" fielded by Gustav Adolf in the 1630s to the horse artillery developed after 1760—by which time anything heavier than 12-pounders (that is, firing 12-pound [5.4-kilogram] balls) was no longer considered suitable for battlefield use. It then became possible to move the guns during the combat, massing them against selected sections of the enemy front as the tactical situation might require. This flexibility, however, was offset by the fact that 18th-century linear formations were almost impossible to turn around. Hence, the really artistic touch consisted of so arranging things as to fall with one's whole force upon one of the enemy's flanks; witness the great victories that Frederick the Great, employing his so-called oblique order, achieved at Rossbach and Leuthen in 1757.

THE FRENCH REVOLUTION

The tactics of the pre-Revolutionary French ancien régime received their final form in the Ordinance of 1791, which reflected the ideas of military writer Jacques Antoine Hippolyte, comte de Guibert; from then until 1831, when the next regulations appeared, formally speaking there was no change. The French Revolution was followed by a short period of tactical improvisation, brought about by the inexperience of the Revolutionary troops, who, unlike their predecessors, were not long-serving regulars but conscripts. However, order was soon restored, and at Jemappes in November 1792 French troops could be observed maneuvering with the best as they pushed away Prussian invaders and won control of Belgium.

As the British general Archibald Percival Wavell observed more than a century later, Napoleon—who served in the French army during the Revolution—was probably a greater strategist than he was a tactician. While he continued the work begun by the Revolution, perhaps his most important tactical innovation consisted of an increased reliance on skirmishers. Previous armies had also made use of skirmishers, but these were mostly irregulars such as the farmers who fired the opening shots in the American Revolution. Since desertion was less of a problem in post-1793 French armies, they could afford to employ regulars in this task. Deploying without any organized formations, skirmishers were permitted to open battles by moving as they saw fit, alternately firing and taking cover. They soon formed as much as one-third of the infantry. Meanwhile, lighter, better-designed artillery (following the system designed by Jean-Baptiste Vaquette de Gribeauval in the last years of the ancien régime) played an ever-increasing role, particularly since

French cuirassiers charging a British square during the Battle of Waterloo on June 18th, 1815. Hulton Archive/Getty Images

the quality of Napoleon's infantry tended to decline after 1808. This permitted "grand batteries" (large artillery units) to be assembled in the midst of battle and fire to be concentrated against selected spots in the enemy front until it was torn to shreds.

These changes apart, the bulk of armies, formed by infantry, continued to deploy much as they had before, and there is no evidence that French methods differed considerably from the rest. Having committed their skirmishers and cannonaded the enemy lines, commanders would form the infantry into one or more columns to launch the assault. Heavy cavalry would be held in reserve to deliver the coup de grâce (deathblow), and this would be followed by light cavalry, which was responsible for pursuit. Perhaps the most effective defensive tactics to counter this system were developed by Britain's duke of Wellington in Spain during the Peninsular War (1808–14). These consisted of drawing up the troops on the reverse side of a ridge, out of the reach of the attacker's artillery, and then allowing the enemy infantry to approach until they could be blasted at almost point-blank range.

TACTICS FROM WATERLOO TO THE BULGE

In many ways, the Battle of Waterloo in 1815—the final battle of the Napoleonic wars, and the end of Napoleon's career—constituted a crucial turning point in the tactics of land warfare. The Industrial Revolution, which began in the 18th century and accelerated in the 19th, introduced a time in which the scale, power, and destructive effect of warfare grew exponentially, culminating in the terrifying spectacle of the atomic bomb.

THE GROWING SCALE OF BATTLE

Before the advent of the industrial age, even though weapons and methods had varied greatly, land battles had essentially been single events, taking up a few square miles and lasting no more than a few hours or a day or two at most. Consisting of formal trials of strength between the main forces of both sides, often enough battles resulted from a kind of tacit mutual consent to commence hostilities. Shifting an army from deep marching columns to thinner and wider fighting formations was a lengthy process; hence, battles very often took on a quasi-ceremonial, paradelike character and were attended by much pomp and circumstance. The short range of weapons—never more than a few hundred yards, usually much less—dictated lateral deployment in order to bring every available man (apart from tactical reserves) into action. Moreover, the means of communication, which had scarcely undergone any change since the dawn of history, imposed definite limits on the length of the fronts that could be controlled by a single commander—three to four miles at most. This in turn meant that the number of troops on each side very rarely exceeded 100,000, a limit that, as mentioned above, had already been reached by Hellenistic times. Indeed, whenever Napoleon brought more than 100,000 men into battle, he tended to lose control over some of them—as happened at Jena, when he forgot about three of the seven corps at his disposal. At Leipzig in 1813, 180,000 French troops faced almost 300,000 Prussians, Russians, Austrians, and Swedes, causing the battle to fall into three separate engagements that were hardly related to one another.

During the 19th century all this was to change, especially as the Industrial Revolution began to make its impact felt on the battlefield after about 1830. Following a century and a half of stagnation, small arms began to undergo rapid technological development. First came percussion caps, then rifled barrels, cylindro-conoidal bullets, breech-loading mechanisms, metal cartridges, and magazines. These improvements permitted tremendous increases in reliability, rate of fire, range, and accuracy—as exemplified by the French Chassepot rifle of 1866, which was sighted to 800 metres (2,600 feet) and was thus theoretically capable of hitting a target at six times the range of the old flintlock musket. Artillery underwent similar development as the old bronze or cast-iron muzzle-loaders gave way to rifled, breech-loading guns made of steel. From the middle of the century, the solid shot and canister that had long formed the principal types of ammunition were replaced by explosive shell, leading to another great increase in lethality and sheer destructive power.

As might be expected, these developments had a profound impact on tactics, even to the point where the very meaning of battle was transformed. Already during Napoleon's time, presenting a solid wall of flesh to the enemy could result in exceedingly heavy casualties. As a result, some of his later battles—Wagram (1809) and Borodino (1812), in particular—were won by mass butchery rather than tactical finesse. Now, however, such methods became positively suicidal. In order to survive on the battlefield, troops, often acting against their officers' wishes, had to discard their brilliant uniforms, lie down, take cover, and disperse. As a result, tightly packed formations disappeared or, in cases when they were retained by obtuse commanders, merely led to horrific casualties. First during the Civil War in the United States, then in Europe, tactical formations began to dissolve. The Prussian chief of staff Helmuth von Moltke expressed concern over the tendency of entire armies to melt into skirmishing lines. The ability of officers to keep their units apart, their men in hand, and their objectives in view declined, if it did not actually disappear. These developments puzzled contemporaries, who came up with the most bizarre ideas as to how to deal with them. In the end, they favoured armies, such as the German one, that adapted to the new circumstances by decentralizing command and making greater use of the individual soldier's initiative.

Insofar as dispersal took place, it caused fronts to grow much longer and less cohesive. From the middle of the 19th century, this tendency was reinforced by the larger number of troops produced by conscription. As battles took up more space, the number of men within a given area declined very sharply. Within each army, fewer troops were actually in action at any moment, giving and receiving fire. One week-long series of engagements during the American U.S. Civil War became known as the Seven Days' Battles. Since modern weapons permitted fighting at longer ranges, gradually a situation was created where the rear areas of armies could be brought under fire just as well as their fronts. Battles, in brief, ceased to be distinct events that could be well defined in time and place and easily identified by crossed swords on a map. During World War I, it became routine for battles to spread over dozens of square miles and last weeks or even months. And, as aircraft became increasingly effective during World War II, they went far to obliterate the distinction between front and rear—another symptom of the changes brought about by modern technology.

The longer that battles lasted, usually the less severe were the casualties produced on any particular day. Throughout the 18th century until the French Revolutionary Wars, armies had fought at the very most three major battles during a campaigning season, which was normally calculated at 180 days. These were bloody affairs, since a few hours of murderous, eye-to-eye combat could easily produce 20, 25, or even 30 percent casualties. However, post-1870 armed forces used their rifled weapons to fire at each other at considerably longer ranges; they also operated in a much more dispersed manner and very seldom brought all or even most of their forces together at a single point. Hence, although over a period of time losses could be just as heavy, they seldom suffered as intensely in a single battle. To suffer casualties in excess of a few percent of strength in one day, as happened to the British at the First Battle of the Somme in 1916, was an exceptional calamity. It was as if, in an instinctive response to the overwhelming power of the new weapons, the fighting became more prolonged but less intense—there being only so much terror that men could stand.

THE POWER OF THE DEFENSE

The last years of the 19th century witnessed the development of automatic weapons in the form of machine guns. Artillery, too, was revolutionized by the addition of recoil mechanisms, which obviated the need to resight the guns after each round and therefore permitted much more rapid fire. As a result the infantry, no longer able to survive the storm of steel sweeping the open terrain, was forced to seek refuge underground. The ineffectiveness of charging cavalry was proved by the immense losses it took during the Crimean and Franco-German wars: unable to follow foot soldiers into underground shelters, it languished and finally disappeared altogether. The tactical defense, rendered invisible by the substitution of smokeless powder for black powder, became much stronger than the offense. This development, the first signs of which could already be seen in the 1850s, dominated the South African War (1899–1902) and the Russo-Japanese War (1904–05)—although most European commanders refused to look facts in the face until the butchery of World War I. During that war, fronts, manned by armies whose troops numbered in the millions, solidified into continuous trench systems that were sometimes hundreds of miles long. Often there were two and even three lines of trenches protected in front by belts of mines and barbed wire hundreds of yards thick. From the rear they were linked to communication trenches, which led into them and allowed reinforcements to arrive without leaving cover.

To overcome a well-entrenched enemy was something that could be achieved, if at all, only by tremendous concentrations of heavy artillery. Directed by forward observers and from balloons and aircraft overlooking the battlefield, artillery fired high explosive, gas, or—ideally, since the two called for different and even contradictory responses—a combination of both. The number of rounds fired could run into the millions; even so, an astute defender needed neither despair nor expose his troops to the physical and psychological effects of a heavy bombardment landing on their dugouts. Instead, leaving only a thin screen to hold the forward line, he could keep his main forces out of the guns' range. As in Wellington's day, the preferred location of such defenses—witness the so-called Hindenburg Line built by the Germans in 1917—was on the reverse slope of a hill or ridge. This denied the enemy observation, complicated his planning, and made it much more difficult for him to register his artillery on target.

In its highest and most developed form, the World War I defensive system consisted of a fortified belt several miles deep. Its main strength was not its continuous trenches but rather its being studded with well-positioned, well-camouflaged strongpoints. So long as the belt held intact, the strongpoints faced forward, bringing fire to bear and acting as observation posts for their own defending artillery. They were, however, also capable of mounting an all-around defense even in the absence of communication with one another and with the rear, thus obstructing the successful attacker as well as delaying and canalizing his progress. Standing ready immediately behind the belt were units (usually the size of regiments, sometimes entire divisions) held in reserve for launching counterattacks. In the German army at any rate, the commanders of such units were often authorized, not to say required, to act on their own initiative without waiting for orders from rear headquarters. The saving of time that was achieved in this way usually permitted local breakthroughs to be quickly repaired, as happened at Cambrai in 1917.

In the face of such defenses, the best-organized attacks were often helpless. Attempts to follow up artillery bombardments by infantry attacking in lines (the method selected by the British at the Somme in 1916) merely led to enormous casualties unequaled in warfare before or since. Later in World War I the Germans, commanded by Erich Ludendorff, developed a new offensive system. The usual daylong and even week-long bombardments were replaced by shorter, more intensive barrages in which gas and high explosive were carefully coordinated and which lasted no more than a few hours. To maintain surprise, no registration rounds were fired, the guns being laid solely by means of mathematical calculation and weather reports. The attacking troops were organized in small, self-contained storming parties. Armed with light machine guns, hand grenades, light mortars, and even some specially designed artillery pieces light enough to be manhandled, they used so-called fire-and-movement tactics. Each subgroup advanced, took cover, and provided the other with covering fire in turn. Like other World War I infantry, the German *Sturmtruppe* suffered greatly from a lack of mobile radio linking them with their own artillery as well as rear headquarters, but, unlike the rest, they were able to overcome this problem to some extent by operating in a decentralized manner, filtering between enemy strongpoints and bypassing resistance in order to penetrate into the rear.

Regarded from a purely tactical point of view, the German methods were very effective. Having proved their worth at Caporetto in 1917, during the great offensives launched in the spring and early summer of 1918 the Germans repeatedly succeeded in driving through British and French defenses. Ultimately, however, they were brought to a halt by the inability of logistic services to follow up over the devastated terrain. Deprived of even the most elementary supplies, the attacking troops were forced to resort to looting and soon lost their cohesion. Sooner or later the breach they made was sealed by the other side's reserves, leaving them stranded in the salient they themselves had created and thus exposed to counterattacks on three sides. It should be added, though, that the World War I offense stood a much better chance of succeeding in theatres other than the Western Front, including, in particular, Poland, Russia, and Palestine. In those theatres modern weapons—especially heavy artillery, which could not be brought up over underdeveloped transportation networks—were often less dense on the ground. Hence attacks could succeed, and in some circumstances even cavalry remained effective.

Another offensive weapon destined to have a great future was the tank. The idea of employing armoured vehicles on the battlefield was not new, dating back at least as far as Leonardo da Vinci (before 1500), but they first appeared on the battlefield in 1916 at the Somme. World War I tanks were either "male" or "female"; that is, they were armed either with cannon up to 75 mm in calibre or else with machine guns. They could drive through wire and cross trenches (sometimes by dropping fascines—bundles of wood—into them), crush or neutralize strongpoints, lay smoke screens, and serve as mobile cover for the infantry to follow. During the last two years of the war they were often employed in all these roles, sometimes

A British tank of the kind that managed to break down the German barbed wire defences at Cambrai, circa 1918. Hulton Archive/Getty Images

with success (as at Amiens in August 1918) and sometimes without. Success often depended on numbers: tanks operating individually or in small groups, it was found, did not have sufficient shock effect. Their armour, only 12 to 16 millimetres (.4 to .6 inches) thick, could be defeated by a determined defender employing field artillery, heavy machine guns, or even special rifles firing heavy ammunition. On the whole, then, early tanks were essentially motorized versions of ancient siege machines. Given their short range, low speed, and general clumsiness, they were suitable for little else.

THE ARMOURED OFFENSIVE

In the decade following World War I, some armies accepted the superiority of defense as a critical characteristic of modern warfare—a train of thought that caused the Maginot Line, an elaborate defensive barrier, to be constructed in northeast France in the 1930s, to guard against German attacks.

Elsewhere, there was a lively debate concerning the best way to break through defensive belts. Aside from air power, two principal solutions were put forward. One, which stressed continued

TANKS AT CAMBRAI

In November–December 1917, after the disastrous Third Battle of Ypres died out in the Flanders mud, the British closed the year's campaign with an operation of some significance for the future. A Tank Corps officer, Col. J.F.C. Fuller, had already suggested a large-scale raid on the front southwest of Cambrai, northern France, where a swarm of tanks, unannounced by any preparatory bombardment, could be released across the rolling downland against the German trenches. This comparatively modest scheme might have been wholly successful if left unchanged, but the British command transformed it; Sir Julian Byng's 3rd Army was actually to try to capture Cambrai and to push on toward Valenciennes. In all, 19 British divisions were assembled, supported by tanks (476 in all, of which about 324 were fighting tanks, the rest being supply and service vehicles) and five horsed cavalry divisions. For the initial attack, eight British divisions were to be launched against three German divisions.

Attacking by complete surprise on November 20, the British tanks ripped through German defenses in depth and took some 7,500 prisoners at low cost in casualties. Bad weather intervened, however, so that the cavalry could not exploit the breakthrough. In addition, all of Byng's tanks had been thrown into the first blow, and adequate infantry reinforcements were not available. By November 29 the offensive had been halted after an advance of about 6 miles (10 km). On November 30 the Germans counterattacked with 20 divisions, and by December 5 the British had been driven back almost to their original positions. Casualties on both sides were about equal—45,000 each. Despite the British failure to exploit the initial success of their tanks, the battle demonstrated that armour was the key to a decision on the Western Front.

development of the light infantry tactics that had achieved partial success in World War I, found particular favour in Germany, where the *Reichswehr* (Germany's post-World War I military organization) was prohibited from developing and deploying heavy weapons and where the chief of staff, Hans von Seeckt, built an elite army that would cut through the defense "like a knife through butter." The other solution, particularly popular in Britain, was armour: improved tanks, operating much like the heavy cavalry of old, were supposed to overcome the defense and restore mobility to the battlefield. There were even visions of armies consisting entirely of tanks.

After 1935 the leading theoreticians reversed their positions. Some of the original proponents of tanks, notably the influential British strategist Basil Liddell Hart, now concluded that the defense had become much the stronger form of war and that armoured offensives would come to grief against a properly organized enemy. In Germany, by contrast, faith in the offensive was never lost, although Adolf Hitler encouraged progressive officers to forsake light infantry and take up tanks—in effect taking the tactical principles pioneered by light infantry in World War I and developing, modifying, and adapting them to armoured warfare. As a result, the *Panzerwaffe* (tank force) was an elite force that grew out of the cavalry rather than the infantry, but it retained many elements of the latter's mode of operations, including an emphasis on interarm cooperation, a decentralized system of command operating within an exceptionally disciplined framework, and a penchant for outflanking and bypassing obstacles rather than confronting them head on.

On a higher level, the Germans saw tanks not as simple siege machines but as fit for playing a strategic role. In World War II, the sequence of the previous war was reversed in that making an initial breach in the enemy's defenses was usually entrusted to the artillery, infantry, and engineers, supported by dive-bombers when the opportunity offered. Once the breach had been made, tanks, accompanied by motorized and later mechanized infantry, poured through. Relying for reconnaissance on the Wehrmacht's (armed forces') ubiquitous motorcycles, they fanned out in the enemy's rear, overran his headquarters, cut his communications, and brought about his collapse by virtue of confusion as much as anything else. To ward off counterattacks against flank and rear, reliance was placed both on the Luftwaffe (Germany's air force) and on excellent antitank artillery (from 1941 some of the latter was mounted on tracked, self-propelled undercarriages, thus creating what were effectively turretless tanks useful both for tank hunting and for close support). To permit all these various troops to cooperate with one another, the Germans added signal troops (they were the first to develop a comprehensive mobile communication system based on two-way radio) as well as a headquarters. Thus, they created the first armoured divisions, which from 1940 became the very symbol of military might.

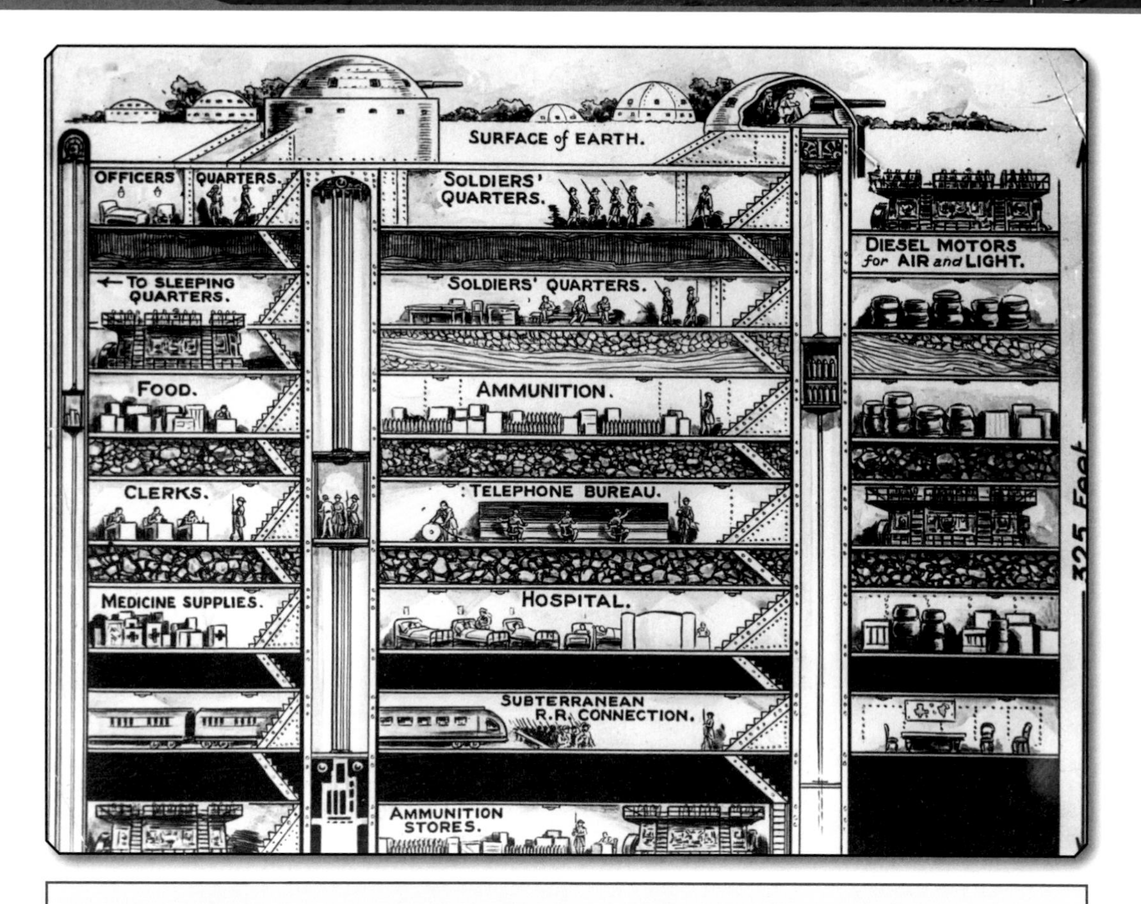

A sectional drawing of the Maginot Line, in 1940, constructed to act as protection against German invasion. Keystone/Hulton Archive/Getty Images

Changes in Command

As armoured tactics developed, the position of the commander as well as the role he played in battle changed. Primitive and ancient commanders, with the partial exception of Roman ones, normally took an active part in the fighting. They and their medieval successors delivered and received blows themselves as a matter of course, with the result that they were sometimes wounded, as was Alexander the Great, or taken prisoner, as was Francis I of France at Pavia in 1525. However, during the second half of the 16th century bureaucratic means of government began to take over from feudalism, and changing social mores no longer required that rulers fight in person. The switch from hand weapons to firearms itself permitted better control, causing commanders to put more emphasis on directing combat and less on participating in it. Increasingly they were to be found not in the midst of their troops but well to the rear, standing on a hill. After about 1650 they could use a "spying glass," or telescope, in order to distinguish their units (newly clothed in uniform) from one another and from the enemy. To communicate their intentions to subordinates they would rely on messengers—and indeed it was in this period that the modern aide-de-camp (military assistant) made his appearance.

An important 19th-century development consisted of electric communication in the form of the telegraph and, later, the telephone. Replacing mounted messengers with the infinitely faster wire made it possible to exercise active command even with armies very far apart and, equally significant, with armies distant from headquarters, located far to the rear. As a result, distances between field commanders—to say nothing of commanders in chief—and their troops tended to increase until they could be measured in miles and even tens of miles. Commanders and their staffs left the field for the office, getting their information by reading reports and bending over maps rather than peering between their horses' ears. After 1860 the old expression *coup d'oeil*, which implied a commander "casting a glance" over the battlefield and making his decision on the spot, was replaced by "estimate of the situation," with its connotation of cooler deliberation. The point was reached when, during World War I, commanders from division level up seldom visited the front; nor would the six-foot-deep trenches, screened by concertina wire in front, have allowed them to take a good look at the enemy even if they had visited it. Moreover, wired communication systems were basically immobile, and efforts to protect them by burying them in the ground tended to make them even more so. In this way they acted as another factor that favoured the defense over the offense.

As commanders came to rely on the wireless communications developed between the world wars, they were able to forsake their headquarters and take to modified tanks, half-tracks, trucks, or even jeeps, which were distinguished from other such vehicles merely by the forest of antennas that they carried. In this way they were able to see the front for themselves and provide leadership at decisive points, all the while keeping in touch with other sectors of the front as well as rear headquarters. In his memoirs, Dwight D. Eisenhower, supreme commander of the Allied forces during World War II, wrote that soldiers usually welcomed his visits because these meant that there was no danger in sight; but other commanders in that conflict, such as Heinz Guderian, Erwin Rommel, George S. Patton, and even Bernard Montgomery (while still merely an army commander) operated in a very different manner from their World War I predecessors. Instead of ensconcing themselves in châteaus, they roamed all over the theatre of war, not seldom taking to the air and covering hundreds of miles in a single day. Regarded from this point of view, radio helped to reverse a secular trend that had been unfolding for centuries, enabling those who knew how to use it to bring about a revolution in command. But for this, modern armoured operations as pioneered in World War II would have been impossible.

LIMITATIONS OF THE TANK

Air forces assisted armoured formations during World War II by providing reconnaissance, interdiction, and close support, as well as putting down airborne troops in front of advancing spearheads when the occasion demanded. Between 1939 and 1942, this method of making war led to brilliant victories equal to any in history. Later, though, it became increasingly clear that there were certain limits to the armoured offensive. Since railways were too inflexible for the purpose, armoured divisions depended on motor convoys for the bulk of their supplies. These convoys themselves made extraordinary demands for fuel, maintenance, and spare parts, with the result that even the most carefully planned, brilliantly led armoured thrusts tended to lose momentum once their spearheads had reached 200 to 250 miles from base. Such an operational reach sufficed to bring down medium-size countries such as Poland and France but not a continent-size country such as the Soviet Union, which was also distinguished by a terrible road system. When the attacker did not enjoy air superiority, as often happened to the Allies before 1942 and to the Germans after that year, the logistic "tails" on which blitzkrieg tactics (a violent surprise offensive by massed air and mechanized ground forces in close coordination) depended proved very vulnerable to attack by fighter-bombers.

Moreover, tanks, originally conceived as offensive instruments, turned out to be at least equally useful on the defense, especially when dug into the ground in "hull-down" positions and deployed with other weapons and field fortifications such as antitank ditches, mines, and barbed wire. Such a combination presented almost insuperable obstacles to the attacker, whose forces would be caught in a maze, cut into penny packets, and lured into killing grounds. Also, as other countries built up their armoured forces in imitation of the Germans, great tank-to-tank battles sometimes took place; but even here the visions of theorists such as J.F.C. Fuller, who had predicted all-tank armies maneuvering against each other like navies at sea, were seldom, if ever, realized. Even in North Africa, with its absolutely open terrain, victory usually went to the side that better knew how to combine armour with other arms such as artillery, antitank artillery, infantry, and, paradoxically, the very engineers whose efforts armour had originally been designed to overcome. From at least 1942, combined-arms warfare became the order of the day, and it remained so for decades to come.

Finally, the tank was not suited for every kind of terrain. Like the cavalry of old, armoured warfare was most effective in broad, open plains like those of northern France, the western Sahara, and southern Russia. In mountainous, forested, swampy, or built-up terrain, the role that tanks could play was necessarily limited, both because of diminished trafficability and because there was insufficient room for them to deploy. Though there were exceptions, often tanks were of no use at all—or else they were reduced to supporting the infantry, as happened in Italy and, later, Korea. Since the tanks' rotating turrets had to absorb the recoil of their guns, these were usually smaller in calibre than ordinary field cannon, so that, employed as artillery, tanks were costly and only moderately effective. Thus, armoured warfare was able to achieve its full potential only in certain theatres. In many others, including Southeast Asia and the Pacific, the role of tanks was more limited, and the old combination of infantry and artillery, now also supported by the air force, usually prevailed.

From Conventional War to Terrorism

No sooner had the mighty state-owned armies of World War II reached their greatest power than the very basis of their supremacy began to disappear. The first development to strike at this basis was nuclear weapons, which threatened to make the very idea of war absurd. Following close after the development of nuclear weapons was the rise of nonstate groups determined to fight for their goals, but not on the same ground as the great engines of state-run warfare.

Nuclear Weapons

On Aug. 6, 1945, the first atomic bomb was dropped on Hiroshima, Japan. From this point, all warfare was destined to be overshadowed by nuclear weapons, devices so powerful as to turn even the mightiest conventional forces into negligible, almost risible, quantities. In theatres where nuclear weapons were present in numbers, such as Europe and Korea, conventional warfare was brought to a dead halt. All attempts to devise ways for fighting in a nuclear environment came to nought, so that the preparations made for it (for example, in the Western doctrine of flexible response) took on a make-believe character and were forced to proceed as if nuclear weapons did not exist. As the strategic nuclear forces of the principal military powers neutralized one another, it was only among—or against—small, unimportant countries that war could be carried on more or less as before. Even then, after about 1970 it became clear that any country in possession of the industrial, scientific, and logistic infrastructure needed to build strong conventional forces would also be able eventually to develop both the bomb and the delivery vehicles it required.

Continued Growth of Military Technology

In spite of its many disadvantages, as listed above, the armoured division continued for several decades following World War II as the very symbol of military might. Immense fortunes were invested in developing, producing, and deploying successive generations of fighting vehicles, especially tanks. On the whole, the weight of tanks, their engine power, and the calibre of their guns trebled between 1940 and 1985, although there were considerable variations in the balancing of armour, armament, and propulsion. The new models incorporated numerous novel features such as stabilized turrets, electronic fire controls, and automatic damage-suppression systems. Nevertheless, in the end tanks remained recognizably what they had been before.

The development of other major weapon systems tended to progress pari passu with that of tanks—and indeed many of them were specifically designed to accompany, assist, or counter them. In order to keep up with their tanks, the most advanced armies became completely motorized. As vehicles for transporting troops, trucks were replaced by armoured personnel carriers; these gave way in turn to armoured fighting vehicles, from which troops could fight without dismounting and some of which were almost as heavy and expensive as tanks. In the rear services, horse-drawn vehicles, which in both the Soviet and German armies had still been in the majority until 1945, disappeared altogether. Consequently, with the bulk of supplies still carried by trucks, the dependence of post-World War II armies on roads was as great as, and possibly greater than, that of their predecessors.

Besides fielding more powerful tanks, troop carriers, and artillery tubes, post-1945 ground forces also introduced entire families of weapons that were absolutely new and unprecedented. Among the earliest were guided antitank missiles, which entered production during the late 1950s but came into their own only with the Arab-Israeli War of October 1973. Short- and medium-range surface-to-surface missiles extended the range of artillery, which was itself increased by providing rounds with added rocket propulsion. Of the missiles, those designed for attacking tanks at short range (two miles or less) proved most effective, forcing armoured divisions to reorganize themselves in order to make possible still closer cooperation between tanks and other arms. Contrary to original hopes, however, they did not bring about either considerable savings in ammunition or relief to logistic systems, the reason being that the standard response to them was to cover every place from which they might be launched with suppressive fire. By and large, the other surface-to-surface missiles were insufficiently accurate, or their warheads too small, to play a decisive role against opposing forces in the field.

In addition to the traditional high explosive, the various new missiles were provided with guidance and homing systems and carried new and powerful warheads such as cluster bomblets and fuel-air explosive. Other missiles were designed for entirely new tasks, such as rapidly scattering large numbers of minelets in front of an advancing opponent. Such tasks presupposed very accurate information on the movements of an opponent who would still be rather far away and, presumably, capable of rapid movement. To provide such information in so-called real time, growing reliance was placed on electronic sensors and unmanned aerial vehicles (UAVs). After becoming familiar in the Vietnam War, where they failed to penetrate the triple-canopy jungle, UAVs became suddenly famous after successful employment by the Israelis in Lebanon in 1982. Launched from mobile platforms and operated by units down to the division level, subsequent generations of UAVs were capable of carrying out surveillance, target acquisition, damage assessment, electronic warfare, and even attacks on the enemy (when provided with homing devices and explosive warheads).

As the jet engine replaced the piston engine in the 1950s and '60s, most aircraft became too fast and unmaneuverable to provide effective close support to ground forces. At the same time, the power of antiaircraft defenses, in the form of missiles and radar-guided, multiple-barrel automatic cannon, increased by leaps and bounds. The Vietnam War and the 1973 Arab-Israeli War demonstrated, each in its own way, the limits of air power in the tactical role, and the 1982 Israeli invasion of Lebanon, in which the Israeli air force won a spectacular victory in the sky without decisively affecting the ground battle, provided even stronger proof. Accordingly, there was a tendency to equip aircraft with long-range guided weapons that would enable them to "stand off" from antiaircraft defenses, and these weapons were used to great effect against Iraq in 1991 in the Persian Gulf War. For close support, increasing reliance was placed on smaller, more agile attack helicopters. The first massive use of helicopters in the air-to-ground role was in Vietnam, where the enemy was generally much too small and dispersed to be effectively tackled by faster craft. Machines armed with guns and missiles specifically designed for "tank busting" entered service during the mid-1970s.

The End of Technological Warfare

Individually, the heavy weapons developed and fielded after 1945 were much more powerful than their predecessors and, thanks to their electronics, capable of hitting faster-moving targets at longer ranges and with greater accuracy. Nevertheless, and in spite of endless talk about the revolutionary changes in warfare brought about by these new arms, the operational art on land stagnated. For 40 years after World War II, the greatest problem confronting Warsaw Pact armies was how to imitate the Wehrmacht and mount a super blitzkrieg aimed at overrunning Europe; simultaneously, the greatest problem confronting the North Atlantic Treaty Organization was how to stop such a blitzkrieg in its tracks. As a result, the great military theorists who pioneered the doctrines of armoured warfare during the 1920s and '30s had no successors of similar stature. Their place was taken by nuclear strategists, whose most important concern was not how to fight a war but how to prevent it from breaking out.

In fact, after 1945 there were only two successful blitzkriegs against worthwhile opponents. The first took place in the Arab-Israeli War of 1967; not accidentally, this saw the use by both sides of many tanks, half-tracks, artillery, and other weapons taken straight out of World War II. The second blitzkrieg was launched at the end of the Persian Gulf War of 1990–91, when the Iraqis, after weeks of saturation bombing, put up so little resistance that only four of the most advanced U.S. tanks were disabled—and none by enemy fire. The October 1973 Arab-Israeli War, by contrast, pointed to the limitations of armoured forces, which suffered high casualties when employed against determined infantry carrying modern antitank weapons or when used as offensive instruments against other armoured forces.

All in all, military forces in the second half of the 20th century were characterized by an unprecedented faith in, and drive for, technology. More and more, land armies deployed their firepower—and their money—in the form of heavy, motorized, crew-operated weapon systems. If only because of their greatly extended ranges, these systems increasingly relied on electronic means for target acquisition, identification, range finding, and aiming. Indeed, the time was to come when the number and quality of electronic gadgets employed by armies became the best possible index of their modernity. However, such devices and their attendant computers operated best of all in simple environments, such as sea and air; in some ways, the most favourable environment of all was outer space, where there was nothing to fight about. Conversely, the more complex the environment, the less reliable and useful modern electronics became, since very often they either gave out the wrong signal or none at all.

The net effect of these factors did not take long to make itself felt. While it became clear that modern armies could inflict enormous attrition on each other, their reliance on long-range, crew-operated, and motorized heavy weapons (and the electronics that these incorporated) also brought about a decrease in those armies' ability to fight opponents that did not resemble themselves—particularly opponents that deliberately chose to operate in complicated terrain, including above all civilian populations and their habitats, communication networks, and means of production. As the Germans in World War II had already learned, in such environments modern weapons, by virtue of their very power, did more harm than good. Panzers (tanks) and dive-bombers could slice through fronts, defeat armies, and overrun countries, but holding those countries down in the face of hit-and-run guerrilla and terrorist attacks was a different matter altogether and could be achieved, if at all, only by old-fashioned infantry.

After 1945 a similar experience was had by virtually every modern army belonging to both developed and developing countries: fighting against organizations other than regular, state-owned armies, they almost always went

British soldiers searching for survivors at the King David Hotel in Jerusalem, Palestine, the British Headquarters, after a massive terrorist bombing by the Zionist military group Irgun in 1946. Fox Photos/Hulton Archive/Getty Images

down to defeat. Technological superiority did not help the French prevail over the Viet Minh in Indochina or the fellaghas in Algeria any more than it enabled the British to defeat the Zionist Irgun Zvai Leumi in Palestine, the Mau Mau in Kenya, or EOKA in Cyprus.

The Soviets in 1979 and the Israelis in 1982 found it easy to overrun Afghanistan and Lebanon, respectively; however, their initial victories proved not so much useless as irrelevant to the final outcome of these wars. The Cubans in Angola (1975–91), the South Africans in Namibia (1975–89), the Indians in Sri Lanka (1987–90), and even the tough Vietnamese in Cambodia (1979–89) all learned the same lesson. In most such cases the insurgents scarcely deployed anything heavier than antitank rockets, machine guns, and light mortars, but often they did not even have those; yet their tactics forced the regular armies to withdraw, sometimes after driving them to the point of complete breakdown, as happened to the Americans in Vietnam. The limitations of conventional forces, their weapon systems, and their methods of making war were highlighted by the fact that conflicts of this kind were far more numerous than conventional ones during the post-1945 period. They also produced by far the most important political results, to say nothing of the number of casualties.

As the 20th century approached its end, there were abundant signs that large-scale, interstate, conventional operations of war had been caught in a vise between nuclear weapons on the one hand and low-intensity operations on the other. In places where nuclear weapons were present—even where the threat was undeclared, as between India and Pakistan or between Israel and its immediate neighbours—such operations were much too dangerous to be attempted. In other places (actually the great majority), where the threat came not from state-owned armies but from other types of organizations with no clear territorial base, conventional warfare was largely useless. Low-intensity warfare had no room for tactics as normally understood and in fact seemed likely to cause them to disappear—that is, to merge with politics and propaganda on the one hand and with terrorism and intimidation on the other. This meant that, even as vast sums continued to be spent on modern conventional weapons and the armies fielding them, the kind of war for which those armies and those weapons were designed seemed to be coming to an end and might, indeed, already have ended.

CHAPTER 3

LOGISTICS

In military science, logistics comprises all the activities of armed-force units in roles supporting combat units, including transport, supply, signal communication, medical aid, and the like.

FUNDAMENTALS OF LOGISTICS

In the conduct of war, war-making activity behind the cutting edge of combat has always defied simple definition. The military vocabulary offers only a few general descriptive terms (such as administration, services, and the French *intendance*), all corrupted by loose usage and none covering the entire area of noncombat activity. All carry additional, though related, meanings that make them ambiguous.

Logistics belongs to this group. Its archaic meaning, the science of computation (from the Greek *logistikos*, "skilled in calculating"), persists in mathematics as the logistic or logarithmic curve but seems unrelated to modern military applications. In the 18th century it crept into French military usage with a variety of meanings, including "strategy" and "philosophy of war." But the first systematic effort to define the word with some precision and to relate it to other elements of war was made by Henri, baron de Jomini (1779–1869), the noted French military thinker and writer. In his *Summary of the Art of War* (1838), Jomini defined logistics as "the practical art of moving armies," by which he evidently meant the whole range of functions involved in moving and sustaining military forces—planning, administration, supply, billeting and encampments, bridge and road building, even reconnaissance and intelligence insofar as they were related to maneuver off the battlefield. In any case,

Jomini was less concerned with the precise boundaries of logistics than with the staff function of coordinating these activities. The word, he said, was derived from the title of the *major général* (or *maréchal*) *des logis* in French 18th-century armies, who, like his Prussian counterpart, the *Quartiermeister*, had originally been responsible for the administrative arrangements for marches, encampments, and troop quarters (*logis*). These functionaries became the equivalent of chiefs of staff to the commanders of the day.

Jomini's discussion of logistics was really an analysis of the functions of the Napoleonic general staff, which he conceived as the commander's right arm, facilitating his decisions and seeing to their execution. The mobility and gargantuan scale of Napoleonic warfare had left the simple old logistics of marches and encampments far behind. The new logistics, said Jomini, had become the science of generals as well as of general staffs, comprising all functions involved in "the execution of the combinations of strategy and tactics."

This broad conception had some validity in Jomini's day. He left an engaging picture of Napoleon, his own logistician, sprawled on the floor of his tent, marking each division's route of march on the map with a pair of dividers. But as staff organization and activity became more complex, along with war itself, the term *logistics* soon lost its association with staff activity and almost disappeared from the military vocabulary. Jomini's great contemporary, the Prussian theorist Carl von Clausewitz, did not share his conception of logistics, which he called "subservient services" that were not part of the conduct of war. Jomini's own influence, which was enormous in his day, was mainly on strategic and tactical thought, particularly in the American Civil War.

American naval scholar Alfred Thayer Mahan, undated photo. U.S. Naval Academy Museum

In the late 1880s the American naval historian Alfred Thayer Mahan introduced logistics into U.S. naval usage and gave it an important role in his theory of sea power. In the decade or so before World War I the navy's concern with the economic foundations of its expansion began to broaden the conception of logistics to encompass industrial mobilization and the war economy. Reflecting this trend, a U.S. marine officer, Lieutenant Colonel Cyrus Thorpe, published his *Pure Logistics* in 1917, arguing that the logical function of logistics, as the third member of the strategy–tactics–logistics trinity, was to provide all the means, human and material, for the conduct of war, including not merely the traditional functions of supply and transportation but also war finance, ship construction, munitions manufacture, and other aspects of war economics.

After World War II the most notable effort to produce a theory of logistics was by a retired rear admiral, Henry E. Eccles, whose *Logistics in the National Defense* appeared in 1959. Expanding Thorpe's trinity to five (strategy, tactics, logistics, intelligence, communications), Eccles developed a conceptual framework that envisaged logistics as the military element in the nation's economy and the economic element in its military operations—that is, as a continuous bridge or chain of interdependent activities linking combat forces with their roots in the national economy. Eccles stressed the tendency of logistic costs to rise (the logistic "snowball") and, echoing Jomini, the essential role of command. Despite its logic and symmetry, however, Eccles' overarching conception of logistics was not widely accepted. Official definitions still vary widely, and most ordinary dictionaries adhere to the traditional "supply, movement, and quartering of troops," but neither has much influence on common usage, which remains stubbornly inconsistent and loose.

Components of Logistics

It is useful to distinguish four basic elements or functions of logistics: supply, transportation, facilities, and services. (A fifth, management or administration, is common to all organized human activity.) All involve the provision of needed commodities or assistance to enable armed forces to live, move, communicate, and fight.

Supply

Supply is the function of providing the material needs of military forces. The supply process embraces all stages in the provision and servicing of military material, including those preceding its acquisition by the military—design and development, manufacture, purchase and procurement, storage, distribution, maintenance, repair, salvage, and disposal. (Transportation is, of course, an essential link in this chain.) The whole process can be divided into four phases: (1) the design-development-production process of creating a finished item, (2) the administrative process by which military agencies acquire finished items, (3) the distribution-servicing processes undergone by military material while "in the service," and (4) the planning-administrative process of balancing supply and demand—that is, the determination of requirements and assets and the planning of production, procurement, and distribution.

U.S. Marines eat their Meals Ready to Eat on April 10, 2003, near Kumayt, Iraq. Keeping soldiers fed is an important part of logistics. Joe Raedle/Getty Images

Military supply has always had the basic aim of providing military forces the material needed to live (food, water, clothing, shelter, medical supplies), to move (vehicles and transport animals, fuel and forage), to communicate (the whole range of communications equipment), and to fight (weapons, defensive armament and materials, and the expendables of missile power and firepower). In all these categories are items, such as clothing, vehicles, and weapons, that are used repeatedly and therefore need to be replaced only when lost, destroyed, or worn out; and materials, such as food, fuel, and ammunition, that are expended or consumed—that is, used only once—and therefore must be continuously or periodically resupplied. From these characteristics are derived the basic classifications of initial issue, replacement, and resupply. The technical classifications of supply vary among countries and services. The British army, for example, recognizes two broad classes: (1) supplies, which include all the expendables except ammunition, and (2) stores, which include ammunition and military hardware. The U.S. Army in World War II and for many years after used five main classifications: (1) subsistence and forage, (2) equipment and other items regularly issued to organizations and individuals, (3) fuels, (4) equipment and materials of irregular issue such as construction materials, and (5) ammunition. These five classes were subsequently expanded to 10 by designating as separate classes certain large categories, such as vehicles, medical material, repair parts, and sales items, which formerly were considered as subclasses.

Historically, food and forage made up most of the bulk and weight of supply until the 20th century, when, with mechanization and air power, fuel displaced forage and became the principal component of supply. However, the demand for food remains unremitting and undeferrable, the one constant of logistics. A man's daily ration makes a small package—7 pounds (3 kg) and often much less. But an army of 50,000 may consume in one month as much as 4,500 tons (4.1 million kg) of food.

Animals require much more. The standard grain and hay ration in the 19th century was about 25 pounds (11.4 kg), and the daily forage of a corps of 10,000 cavalry weighed as much (allowing for remounts) as the food for 60,000 men. Forage requirements tended, moreover, to be self-generating, since the animals needed to transport it also had to be fed. The number of animals accompanying an army varied widely. Napoleon's ideal, which he himself never attained, was a supply train of only 500 wagons in an army of 40,000; with a corps of 7,000 cavalry, this would amount to about 10,000 animals exclusive of remounts and spare draft animals. Northern armies in the American Civil War commonly numbered half as many animals as soldiers. A force of 50,000 men might thus require more than 300 tons (272,000 kg) of forage daily. This was more than twice the weight of gasoline that an equivalent force of three World War II infantry divisions, using motor vehicles exclusively, needed to operate for the same length of time. In the latter case, moreover, fuel requirements diminished markedly when an army was not moving, whereas the premechanized force had to feed its animals whether moving or not. It was the immense forage requirements of premechanized armies, more than any other single factor, that restricted warfare before the 20th century so generally to seasons and climates when animals and men could subsist mainly on the countryside.

In 20th-century warfare the expendables of movement included fuel for rail and water transport as well as for motor vehicles, and also the immense fuel requirements of modern air power. In World War II, without counting transoceanic shipment, fuel made up half the resupply and replacement needs of U.S. forces in Europe. Technologically advanced warfare has, in fact, vastly increased fuel consumption both absolutely and relatively to other supply needs. The continued development of mechanization and air power has increased by one and one-half times the fuel requirements of large-scale conventional military operations typical of World

War II. Food, by contrast, is a small and diminishing fraction of the total burden.

Before the 20th century, equipment replacement and ammunition resupply were a relatively small part of an army's needs. Missile power before the gunpowder era was limited by the difficulty of bringing missiles in quantity to the battlefield. For the first five centuries of the gunpowder era the provision of ammunition was not a major logistic problem. Not until the use of field artillery on a large scale in the late 18th century, and the development of quick-firing shoulder arms in the 19th, did ammunition begin to constitute a substantial part of resupply needs. As late as 1864, in the Atlanta campaign of the American Civil War, the Union army's average daily ammunition requirements amounted to only one pound (0.45 kg) per man, as against three pounds for rations; Confederate forces in that war were reported to expend, on the average, only half a cartridge per man per day.

The great increase in firepower in the 20th century upset the historic ratios. In World War II the average ammunition requirements of Western forces in combat zones were 12 percent of total needs. In the mainly positional Korean War, ammunition expenditures climbed higher, and a late-1980s U.S. Army planning factor rated ammunition requirements as more than one-quarter of total supply. Material replacement needs have also mounted in absolute terms; the great tank battles of World War II and of the Arab-Israeli Wars of 1967 and 1973 involved the destruction of hundreds of tanks within a few days. But as a percentage of total supply, replacement of material losses is a declining factor.

TRANSPORTATION

Before the development of steam propulsion, armies depended for mobility on the muscles of men and animals and the force of the wind. On land they used men and animals to haul and carry; on water they used oar-driven and sail-propelled vessels. Among these various modes the balance of advantage was often delicate. A force moving by water was vulnerable to storm and enemy attack; navigation was an uncertain art; transports were expensive and of limited capacity. Large expeditions could be undertaken only by wealthy states or seafaring peoples, such as the Scandinavians of the 8th and 9th centuries, who combined the roles of mariner and warrior. Seaborne armies were rarely strong enough to overcome a resolute land-based foe.

On the other hand, armies have usually been able to move faster and with a better chance of avoiding enemy detection by water than by land. Shipment of bulky freight is cheaper and safer by river than by road, and good roads are rare in military history. In the 19th and 20th centuries the revolution in ship design and propulsion made water travel largely independent of wind and weather, permitting the overseas movement and support of larger forces than ever before. After the mid-19th century, however, more and better roads and, above all, railroads began to offset the historic advantages of water transportation to some degree. In the 20th century motor vehicles and more road building extended the conquest of rough terrain. The airplane finally freed military movement, for modest forces and limited cargo, from bondage to earth altogether. Yet the costs of mobility on land—in equipment, materials, and energy—remain high, and large military movements are still confined to narrow ribbons of rail and road, which in many parts of the world are still rare or lacking.

On land the soldier himself has been the basic burden carrier of armies. As a matter of simple economy, he represents large carrying capacity at no extra cost. His equivalent, in an army of 50,000 in the preindustrial era, would be 1,875 wagons drawn by 11,250 horses or mules, which might need additional wagons and animals to haul forage. A difference of only 5 pounds (2.3 kg) in the soldier's load could add or subtract a requirement for 125 wagons and 750 animals. Since the days of the Roman legion, the soldier has had to carry, on the average, about 55 or 60 pounds (25 or 27 kg). The ratio between weapons and other items in the soldier's load has varied widely, but the modern soldier has relegated most of his food to vehicle transport while still carrying a heavy burden of weapons and ammunition. Since World War II, however, some armies have made drastic reductions in the combat load.

Before the age of mechanization, the soldier's carrying capacity was usually supplemented by additional carriers and haulers, human and animal. Each had advantages. A team of six horses ate about as much as 30 to 40 men, but the men could carry more on their backs than the horses could haul and considerably more than the horses could carry. Men could negotiate rougher terrain, and they required less care. On the other hand, loads placed on men had to be distributed in small packages, and men proved less efficient than animals when teamed to haul heavy and bulky loads. The horse and mule, however, have less strength and stamina, though more agility, than the ox, history's primary beast of burden. In many parts of the world, motor transport still has not displaced human and animal carriers and haulers in the movement of military supply.

FACILITIES

The provision of military facilities, as distinct from fortification, did not become a large and complex sphere of logistic activity until the transformation of warfare in the industrial era. In that transformation the traditional function of providing nightly lodgings or winter quarters for the troops dwindled to relative insignificance in the mushrooming infrastructure of fixed and temporary installations that became part of the military establishments of the major powers. Modern armies, navies, and air forces own and operate factories, arsenals, laboratories, power plants, railroads, shipyards, airports, warehouses, supermarkets, office buildings, hotels, hospitals, homes for the aged, schools, colleges, and many other types of structures used by advanced societies—as well as barracks, the original military facility. They are among the world's great landowners. The management of all this improved real estate is one of the largest areas of modern logistic administration.

Services

Services may be defined as activities designed to enable personnel or material to perform more effectively. Usage recognizes no clear distinction between logistic and nonlogistic services, but a somewhat blurred one has grown out of the traditional and opprobrious identification of logistics with noncombat rear-area activities. Thus, intelligence and communications personnel and combat engineers in the U.S. Army have long claimed the label of "combat support" as distinct from the "service support" functions of supply, transportation, hospitalization and evacuation, military justice and discipline, custody of prisoners of war, civil affairs, personnel administration, and nontactical construction (performed by "construction" engineers). Training of combat troops is hardly ever considered a logistic service, whereas training of service troops sometimes is. Usage does not, however, always assign "service support" to logistics. Personnel administration is an old, institutionalized sector of the military establishment, and personnel administrators tend to reject the logistics label. Personnel services (medical, spiritual, educational, financial) are more heterogeneous and have varied origins; most definitions of logistics include them.

Most service activities, logistic and nonlogistic, are of recent origin and, as organized specialities, are peculiar to the military establishments of advanced nations. Over the long haul of military history, the services considered necessary to keep armed forces in fighting trim were generally of a rudimentary character. From the earliest times, however, they posed a serious logistic problem. To armies and their lines of communication they added numbers of people who did not, as a primary function, belong to the fighting force and who, if not properly organized, might weaken its capacity to fight. Soldiers seldom possessed the technical skills required to perform any but the simplest services; sometimes, as members of a warrior elite, they were prohibited by social prerogative from performing them. A classic feature of armies, consequently, has been its long train of noncombatants, often far outnumbering the fighting men.

Logistic services also added to the baggage of armies a growing burden of specialized equipment, tools, and materials needed for the performance of the services. Services tended to generate more services: service equipment itself had to be serviced, sometimes by additional technicians, and service personnel themselves required services. Logistic services thus meant more people to be fed, clothed, and sheltered and more people and baggage to be transported. What the British call the "administrative tail" is as old as military history.

Special Features of Naval Logistics

From early times, the substantial carrying capacity of the warship made it an indispensable element in its own logistic support, particularly in the era before steam power eliminated the problem of covering long distances between ports. (Oar-driven warships, such as the Greek trireme, sacrificed this feature in order to maximize fighting power.) For centuries the most critical item of supply was water, which sailing ships found difficult to carry in sufficient quantities and to keep potable for long voyages. Food was somewhat less of a problem, except for its notoriously poor quality in the days before refrigeration, the sealed container, and sterilization.

During the long reign of the sailing ship, the absence of a fuel requirement was a major factor in the superior mobility of fleets over armies. The shift to steam was, in a sense, a return to the principle of self-contained propulsion earlier embodied in the oar-driven ship. The gain in control was of course an immeasurable improvement for the long haul, but for a time the inordinate amount of space that had to be allocated to carry wood or coal seriously inhibited the usefulness of early warships. Eventually the maritime nations established networks of coaling stations, which became part of the fabric of empire in the late 19th century. The shift to oil a few years before World War I involved a major dislocation in naval logistics and changed the stakes of imperial competition.

For modern navies the importance of bases goes far beyond the need for periodic replenishment of fuel, although this remains essential. Ships must be repaired, overhauled, and resupplied with ammunition and food; and, an ancient requirement, the crews must be given shore leave. Within limits, these needs can be filled by specialized auxiliary ships either accompanying naval forces at sea or stationed at predetermined rendezvous points. Naval operations in World War II saw a proliferation of these auxiliary vessels; in 1945 only 29 percent of the U.S. Navy consisted of purely fighting ships. By using auxiliaries and by rotating ships and personnel, modern fleets can remain at sea indefinitely, especially if not engaged in combat. U.S. fleets in the Mediterranean and far Pacific have done so for years, although the feat is less impressive than that of the British admiral Lord Nelson's fleet, which lay off Toulon, France, continuously, without rotation, for 18 months from 1803 to 1805, in the war against Napoleon. With nuclear propulsion, thus far applied only to submarines and a handful of large warships, the basic logistic function of replenishing fuel may eventually disappear. But that day will be long in coming, and the other functions of naval logistics will remain.

Power Versus Movement

The potential effectiveness of a military force derives from three attributes: fighting power, mobility, and range of movement. Which of these attributes is stressed depends on the commander's objectives and strategy, but all must compete for available logistic support. Three methods have been used, in combination, in providing this support for forces in the field: self-containment, local supply, and supply from bases.

SELF-CONTAINMENT

The idea of complete independence from external sources of supply—the hard-hitting, self-contained "flying column"—has always been alluring but has seldom fully materialized. Self-containment in weapons, equipment, and missiles or ammunition was common enough before the great expansion of firepower and resupply requirements in the last century. But few military forces have been able to operate for long or move far without frequent resupply of food and forage or fuel.

Self-containment is the least economical of all methods of supply. Accompanying transport is fully employed only at the beginning of the movement, serving thereafter as a rolling warehouse that is progressively depleted as the force moves. Fast-moving, self-contained forces typically left a trail of abandoned vehicles and dead animals. The basic trade-off in self-containment is between the speed gained by avoiding delays and detours for foraging and the speed lost by dragging a large baggage train. When Hannibal crossed the Alps into northern Italy in 218 BCE, he bypassed the Roman army guarding the easier coastal route; but his movement through the mountain passes was painfully slow, and he lost almost half his force to cold, disease, and hostile tribes along the way.

LOCAL SUPPLY

Until the 20th century, armies commonly lived off the country and, in enemy territory, from captured stores. In fertile regions an army could usually provision itself at low cost in transport and without sacrificing fighting power or range; when efficiently organized, local supply even permitted a high degree of mobility. Normally, however, an army living off the country tended to straggle and to load itself down with loot. If it moved too slowly or was pinned down, it might sweep the region bare and starve. In winter, in deserts and mountains, or in thinly populated areas, local supply offered meagre fare. And a hostile population, as Napoleon discovered in Russia and Spain, could bring disaster to an army that had to scrounge for its food. (British forces in the American colonies during the Revolution had to draw most of their supplies from overseas.) Animals, in any case, almost always had to shift for themselves. Cattle driven with an army could transform forage into food, a supply technique as ancient as the Bible and still common in the 19th century. Unwieldy and slow-moving though it was, the accompanying herd had the great merit of transporting itself and dwindling as it was consumed.

When mechanized transport replaced animals, one of the great continuities of military history was broken. Mechanized

Hannibal Crosses the Alps

In 219 BCE the great Carthaginian general Hannibal captured Saguntum (Sagunto), an independent city on the east coast of Spain with which Rome had an understanding (though perhaps not an actual treaty) of "friendship." Rome demanded Hannibal's withdrawal, but Carthage refused to recall him, and Rome declared war (the Second Punic War). Because Rome controlled the sea, Hannibal led his army overland through Spain and across the Pyrenees into Gaul. He crossed the Alps into the upper Po River valley of Italy either by the Col de Grimone or the Col de Cabre and then through the basin of the Durance, or else by the Montgenèvre or Mont Cenis pass.

Some details of Hannibal's crossing of the Alps have been preserved. At first danger came from the Allobroges, who attacked the rear of Hannibal's column. (Along the middle stages of the route, other Celtic groups attacked the baggage animals and rolled heavy stones down from the heights on the enfilade below, thus causing both men and animals to panic and lose their footings on the precipitous paths. Hannibal took countermeasures, but these involved him in heavy losses in men.) On the third day he captured a Gallic town and from its stores provided the army with rations for two or three days. Harassed by the daytime attentions of the Gauls from the heights and mistrusting the loyalty of his Gallic guides, Hannibal bivouacked on a large bare rock to cover the passage by night of his horses and pack animals in the gorge below. Snow was falling on the summit of the pass, making the descent even more treacherous. Upon the hardened ice of the previous year's fall, the soldiers and animals alike slid and foundered in the fresh snow. A landslide blocked the narrow track, and the army was held up for one day while it was cleared.

Finally, on the 15th day, after a journey of five months, with 20,000 infantry, 6,000 cavalry, and only one of the original 37 elephants (the sole Asian elephant among 36 African), Hannibal descended into Italy, having surmounted the difficulties of climate and terrain, the guerrilla tactics of inaccessible tribes, and the major difficulty of commanding a body of men diverse in race and language under conditions to which they were ill-fitted. Descending into the Po valley, the territory of the hostile Taurini, Hannibal stormed their chief town (modern Turin).

armies can operate in winter and desert areas as long as they have fuel; when that runs out, they grind to a halt. Until fuel can be compressed into small capsules (as, in a sense, atomic energy is) or, like forage, be gathered along the way, the door to both self-containment and local supply will remain closed.

Supply from Bases

The alternative to self-containment and local supply is continuous or periodic resupply and replacement from stores prestocked at bases or other accessible points. Supply from bases involves three serious disadvantages. First, supply routes are often vulnerable to attack. Second, an army shackled to its bases lacks flexibility and moves slowly—even more slowly as it advances. Finally, the transportation costs of maintaining a flow of supply over substantial distances are heavy and, beyond a point, prohibitive. The reason is twofold; first, because the transport of the supply train must operate a continuous shuttle—that is, for each day's travel time, two vehicles are needed to deliver a single load—and, second, because additional food and forage or fuel must be provided for the personnel, animals, or vehicles of the train itself. In the era of animal-drawn transport this multiplier factor set practical limits to the operating radius of an army, which the American Civil War general William T. Sherman fixed at about 100 miles (160 km), or five days' march, from its base. The critical limitation was the provision of forage, the bulkiest supply item. For an army operating at any considerable distance from its bases, the in-transit forage requirements of its shuttling supply train, if supplied entirely from bases, would saturate any amount of transport, leaving none to supply the fighting force. Since pre-mechanized armies usually found some local forage and food, supply from bases, in combination with local supply and an accompanying train, was the normal method, but Sherman's 100 miles was seldom exceeded.

With modern mechanized transport the theoretical maximum operating radius is so great that other limitations come into play. Nevertheless, the in-transit fuel needed to supply a force from distant bases adds major increments of transport cost, especially under conditions (e.g., poor roads) that reduce speed or increase fuel consumption. It can also severely limit the speed of an advancing mechanized force, as shown by the bogdown of the U.S. 3rd Army's drive across France in the summer of 1944 for lack of fuel.

HISTORICAL DEVELOPMENT

The universal principles of supplying war have been applied in three major periods: the long period of history when war was powered by human and animal muscle; the approximately 100 years from the mid-19th century through World War II, when industrial might changed warfare profoundly; and the modern nuclear age, when weapons of mass destruction and technological change have removed certain age-old problems of logistics and created new ones.

LOGISTIC SYSTEMS BEFORE 1850

In ancient history the combination of local supply for food and forage and self-containment in hardware and services appears often as the logistic basis for operations by forces of moderate size. Some of these operations are familiar to many a schoolchild—the long campaign of Alexander the Great from Macedonia to the Indus, the saga of Xenophon's Ten Thousand, Hannibal's campaigns in Italy. The larger armies of ancient times—like the Persian invaders of Greece in 480 BCE—seem to have been supplied by depots and magazines along the route of march. The Roman legion combined all three methods of supply in a marvelously flexible system. The legion's ability to march fast and far owed much to superb roads and an efficiently organized supply train, which included mobile repair shops and a service corps of engineers, artificers, armourers, and other technicians. Supplies were requisitioned from local authorities and stored in fortified depots; labour and animals were drafted as required. When necessary, the legion could carry in its train and on the backs of its soldiers up to 30 days' supply of provisions. In the First Punic War against Carthage (264–241 BCE), a Roman army marched an average of 16 miles (26 km) a day for four weeks.

One of the most efficient logistic systems ever known was that of the Mongol cavalry armies of the 13th century. Its basis was austerity, discipline, careful planning, and organization. In normal movements the Mongol armies divided into several corps and spread widely over the country, accompanied by trains of baggage carts, pack animals, and herds of cattle. Routes and campsites were selected for accessibility to good grazing and food crops; food and forage were stored in advance along the routes of march. On entering enemy country, the army abandoned its baggage and herds, divided into widely separated columns, and converged upon the unprepared foe at great speed from several directions. In one such approach march a Mongol army covered 180 miles (290 km) in three days. Commissariat, remount, and transport services were carefully organized. The tough and seasoned Mongol warrior could subsist almost indefinitely on dried meat and curds, supplemented by occasional game; when in straits, he might drain a little blood from a vein in his mount's neck. Every man had a string of ponies; baggage was held to a minimum, and equipment was standardized and light.

In the early 17th century, King Gustav II Adolf of Sweden and Prince Maurice of Nassau, the military hero of the Netherlands, briefly restored to European warfare a measure of mobility not seen since the days of the Roman legion. This period saw a marked increase in the size of armies; Gustav and his adversaries mustered forces as large as 100,000, Louis XIV of France late in the century even more. Armies of this size had to keep on the move to avoid starving; as long as they did so, in fertile country they could usually support themselves without bases, even with their customary huge noncombatant "tail." Logistic organization improved, and Gustav also reduced his artillery train and the size of guns. In the Thirty Years' War (1618–48) strategy tended to become an appendage of logistics as armies, wherever possible, moved and supplied themselves along rivers exploiting the economies of water transportation, and operated in rich food-producing regions.

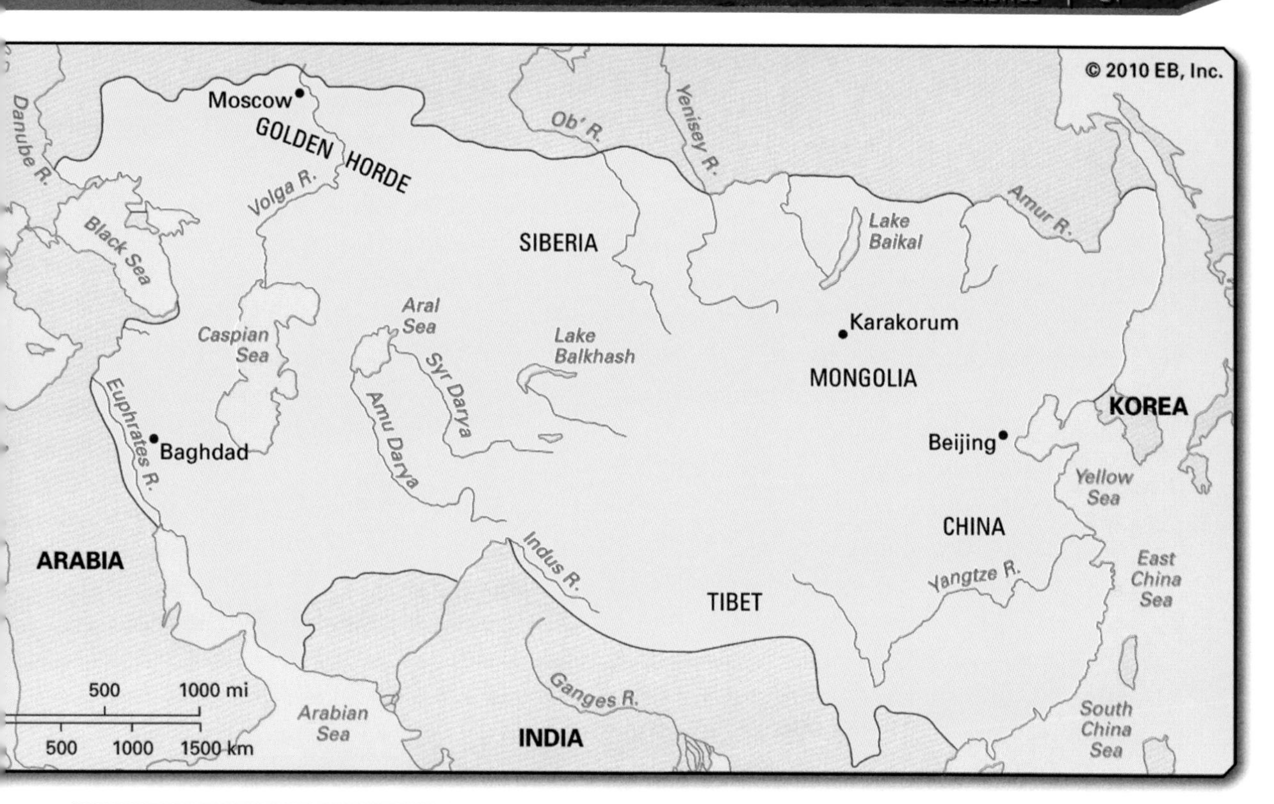

The Mongol Empire of the 13th century.

After the Thirty Years' War, European warfare became more sluggish and formalized, with limited objectives and an elaborate logistics that sacrificed both range and mobility. The new science of fortification made towns almost impregnable while enhancing their strategic value, making 18th-century warfare more an affair of sieges than of battles. Two logistic innovations were notable: the magazine, a strategically located prestocked depot, usually established to support an army conducting a siege; and its smaller, mobile version, the rolling magazine, which carried a few days' supply for an army on the march. Secure lines of communication became vital, and whole armies were deployed to protect them. The increasing size of armies and of artillery and baggage trains placed heavier burdens on transport. Also, a revulsion against the depredations and inhumanity of the 17th-century religious wars resulted in curbs on looting and burning and in regulated requisitioning or purchase of provisions from local authorities. Because of the high cost of mercenary soldiery, commanders tended to avoid battles, and campaigns tended to become sluggish maneuvers aimed at threatening or defending bases and lines of communication. "The masterpiece of a successful general," Frederick the Great remarked, "is to starve his enemy."

The era of the French Revolution and the Napoleonic domination of Europe (1789–1815) brought back both mobility and range of movement to European warfare, along with an immense further increase in the size of armies. Abandoning the siege warfare of the 18th century, Napoleonic strategy stressed swift offensives aimed at smashing the enemy's main force in a few decisive battles. The logistic system inherited from the Old Regime proved surprisingly adaptable to the new scale and pace of operations. Organization was made more efficient, baggage trains were pared down and some of their load shifted to the soldier's back, and much of the noncombatant tail was eliminated. The artillery train was increased, and the rolling magazine was used as the occasion demanded. The heavily burdened citizen-soldier marched faster and farther than his mercenary predecessor. In densely populated and fertile regions, moving armies continued to subsist, by purchase and requisition, on the countryside through which they marched, spreading out over parallel roads, each corps foraging to one side only. Even so, the numbers involved dictated greater dependence on magazines.

Napoleon made relatively few logistic innovations. He militarized some services formerly performed by contractors and civilian personnel, but the supply service (*intendance*) remained civilian though under military control. A significant change was the establishment in 1807 of a fully militarized train service to operate over part of the line of communication; this was divided into sections that were each serviced by a complement of shuttling wagons—foreshadowing the staged resupply system of the 20th century. The 600-mile (1,000-km) advance of Napoleon's Grande Armée of 600,000 men into Russia in 1812 involved logistic preparations on an unprecedented scale. Despite extensive sabotage by the Russian peasantry, the system brought the army victorious to Moscow.

Logistics in the Industrial Era

Between the mid-19th and the mid-20th centuries the conditions and methods of logistics were transformed by a fundamental change in the tools and modes of making war—perhaps the most fundamental change since the beginning of organized warfare. The revolution had four facets: (1) the mobilization of mass armies; (2) a revolution in weapons technology involving a phenomenal increase in firepower; (3) an economic revolution that provided the means to feed, arm, and transport mass armies; and (4) a revolution in the techniques of management and organization, which enabled nations to operate their military establishments more effectively than ever before.

Mass Mobilization

These interrelated developments did not occur all at once. Armies of unprecedented size had appeared in the later years of the Napoleonic Wars. But for almost a century after 1815, the world saw no comparable mobilization of manpower except in the American Civil War. Meanwhile, the growth of population (in Europe, from 180 million in 1800 to 490 million in 1914) was creating a huge reservoir of manpower. By the end of the 19th century most nations were building large standing armies backed by even larger partially trained reserves. In the world wars of the 20th century the major powers mobilized armed forces numbering millions.

The revolution in weapons had started earlier but accelerated after about 1830. By the 1850s and '60s the rifled percussion musket, rifled and breech-loading artillery, large-calibre ordnance, and steam-propelled armoured warships were all coming into general use. The revolution proceeded with gathering momentum thereafter, but it remained for mass armies in the 20th century to realize its full potential for destruction.

By the mid-19th century the Industrial Revolution had already given Great Britain, France, and the United States the capacity to produce munitions, food, transport, and many other items in quantities no commissary or quartermaster had ever dreamed of. But except in the Northern states during the American Civil War, the wars of the 19th century hardly scratched the surface of the existing war-making potential. The nature of international rivalries of the period tended to limit war objectives and the mobilization of latent military power. Only in the crucible of World War I, at the cost of colossal blunders and wasted effort, did nations begin to learn the techniques of "total" war. Long before 1914, however, new instruments and techniques of logistics were emerging.

Transportation and Communication

The railroad, the steamship, and the telegraph had a profound impact on logistic method during the last half of the 19th century. Beginning with the Crimean War (1854–56), telegraphic communication became an indispensable tool of command, intelligence, and operational coordination, particularly in controlling rail traffic. In the 20th century it yielded to more efficient forms of electronic communication—the telephone, radio, radar, television, telephotography, and the high-speed computer.

Railroads spread rapidly over western and central Europe and the eastern United States between 1850 and 1860. They were used—mainly for troop movements—in the suppression of central European revolutions in 1848–49, on a considerable scale in the Italian War of 1859, and extensively in the American Civil War, where they also demonstrated their capacity for long hauls of bulky freight in sustaining the forward movement of armies. In Europe, from 1859 on, railroads shaped the war plans of all the general staffs, the central features of which were the rapid mobilization and concentration of troops on a threatened frontier at the outbreak of war. In 1870, at the outset of the Franco-German War, the German states were able to concentrate 550,000 troops, 150,000 horses, and 6,000 pieces of artillery on the French border in 21 days. Germany's recognized efficiency in mobilizing influenced the war plans of all the European powers in 1914. In both world wars Germany's railroads enabled it to shift troops rapidly between the Eastern and Western fronts.

Steam propulsion and iron ship construction also introduced new logistic capabilities into warfare in the 19th century. Steamships moved troops and supplies in support of U.S. forces in the Mexican War of 1846–48 and of British and French armies in the Crimea. River steamboats played an indispensable role in the American Civil War.

The complement of the railroad was the powered vehicle that could travel on ordinary roads and even unprepared surfaces, within the operating zones of armies forward of railheads. This was a 20th-century development, a combination of the internal-combustion engine, the pneumatic tire, and the endless track. Motor transport was used on an increasing scale in both world wars, although animal-drawn transport and railroads still dominated land movement. Another innovation was the pipeline, used to move water in the Palestine campaign of World War I and extensively in World War II to move oil and gasoline to storage points near the combat zones. More revolutionary was the development of large-scale air transportation. In World War II, units as large as a division were carried in one movement by air over and behind enemy lines and resupplied by the same means. Cargo aircraft maintained an airlift for more than three years from bases in India across the Himalayas into China; during the last eight months of operation it averaged more than 50,000 tons per month. But the fuel costs of such an operation were exorbitant. Air transportation remained primarily a means of emergency movement when speed was an overriding consideration.

The Growth in Quantity

The most conspicuous logistic phenomenon of the great 20th-century wars was the enormous quantity of material used and consumed. One cause was the growth of firepower, which was partly a matter of increased rapidity of fire of individual weapons, partly a higher ratio of weapons to men—both multiplied by the vast numbers of troops now mobilized. An American Civil War infantry division of 3,000 to 5,000 men had an artillery complement of up to 24 pieces; its World War II counterpart, numbering about 15,000 men, had 328 artillery pieces, all capable of firing heavier projectiles far more rapidly. A World War II armoured division had nearly 1,000 pieces of artillery. Twentieth-century infantrymen, moreover, were armed with semiautomatic and automatic weapons.

The upward curve of firepower was reflected in the immense amounts of ammunition required in large-scale operations. Artillery fire in the Franco-German War and in the Russo-Japanese War (1904–05), for example, showed a marked increase over that in the American Civil War. But World War I unleashed a firepower hardly hinted at in earlier conflicts. For the preliminary bombardment (lasting one week) in the First Battle of the Somme in 1916, British artillery was provided 23,000 tons of projectiles; 100 years earlier, Napoleon's gunners at Waterloo had about 100 tons. In World War II the United States procured only about four times as many small arms as it had in the Civil War but 43 times as much small-arms ammunition. (To the ammunition expenditures in World War II were added, moreover, the immense tonnages of explosives used in air bombardment.) The Confederacy fought through the four years of the Civil War on something like 5,000 or 6,000 tons of gunpowder, whereas U.S. factories in one average month during World War I turned out almost four times this quantity of smokeless powder. Again, in one year of World War II, seven million tons of steel went into the manufacture of tanks and trucks for the U.S. Army, four million tons into artillery ammunition, one million tons into artillery, and 1.5 million tons into small arms—as contrasted with less than one million tons of pig iron used by the entire economy of the Northern states during one year of the Civil War.

With quantitative growth went a parallel growth in the complexity of military equipment. The U.S. Army in World War II used about 60 major types of artillery above .60-inch calibre; for 20 different calibres of cannon there were about 270 types and sizes of shells. The list of military items procured for U.S. Army ground forces added up to almost 900,000, each of which contained many separate parts—as many as 25,000 for some antiaircraft guns. To convert and expand a nation's peacetime industry to the production of such an arsenal posed staggering technical problems. Manufacturers of automobiles, refrigerators, soap, soft drinks, bed springs, toys, shirts, and microscopes had to learn how to make guns, gun carriages, recoil mechanisms, and ammunition.

STAGED RESUPPLY

Long before mechanization relegated local supply to a minor role in logistics, growing supply requirements were making armies more dependent on supply from bases. The *Etappen* system of the Prussian army in 1866 resembled the Napoleonic train service of 1807. Behind each army corps trailed a lengthening series of shuttling wagon trains moving up supplies through a chain of magazines extending back to a railhead. A small train accompanied the troops, carrying a basic load of ammunition, rations, and baggage; each soldier also carried additional ammunition and three days' emergency rations. The system was geared to a steady, slow advance on a rigid schedule and a predetermined route.

October 1916: British artillery men transport a gun through the Somme during World War I. Topical Press Agency/Hulton Archive/Getty Images

Before the advent of mechanization half a century later, the system did not work well, since the shuttling wagon trains were unable to keep up with a rapid advance. In both the Franco-German War and the German invasion of France in 1914, German forces outran their trains and had to live off the French countryside, one of the richest agricultural regions in Europe. In the latter campaign, however, the Germans' tiny motor transport corps played a vital role in supplying ammunition for the opening battles. In subsequent operations on the Western Front, the immobility of the opposing forces provided an ideal environment for the staged resupply system, reversing the ancient rule that a "sitting" army must starve. On the other hand, many offensives on that front bogged down, after gaining only a few miles, through failure to move up quickly the quantities of fuel, ammunition, and supplies needed to maintain momentum.

The staged resupply system, in practice, did not precisely resemble either a pipeline or a series of conveyor belts maintaining a continuous flow from ultimate source to consumer. Reserves were stocked as far forward as was safe and practicable, permitting a regular supply of food and fuel and an immediate provision of ammunition, equipment, and services as needed. Before a major operation, large reserves had to be accumulated close behind the front; the two-year Allied buildup in the British Isles before the Normandy invasion of 1944, for example, involved the shipment of 16 million tons of cargo across the Atlantic. After the invasion, behind the armies on the Continent spread the rear-area administrative zone, a vast complex of depots, traffic regulating points, railway marshaling yards, troop cantonments, rest areas, repair shops, artillery and tank parks, oil and gasoline storage areas, air bases, and headquarters—through which ran the lines of supply stretching back to ultimate sources.

In the Pacific, the administrative zone covered vast reaches of ocean and clusters of islands. Communication and movement in this theatre depended largely on shipping, supplemented by aircraft, and one of the major logistic problems was moving forward bases and reserves as the fighting forces advanced. Supply ships often sailed all the way from the U.S. West Coast, bypassing intermediate bases, to forward areas where they were held as floating warehouses until their cargoes were exhausted.

In a real sense, the basic logistic tools of land operations in World War II were the railroad, the motor truck, and, carried over from the premechanized era, the horse-drawn wagon. Motor transport, when available, served to move forward the mountains of material brought to railheads by the railroads—a feat that, as the late 19th-century wars and World War I had shown, could not be done by horse-drawn vehicles rapidly enough to sustain fast-moving forces. When supplied by motor transport, mechanized armies, particularly in the European theatre, achieved a mobility and striking power never before seen. Paradoxically, Germany, which dominated operations in this theatre until late in the war, suffered from a severe shortage of motor transport and rolling stock, only partially made good by levies on conquered nations. The Wehrmacht that invaded the Soviet Union in 1941 consisted mainly of slow-moving infantry divisions supplied by horse-drawn wagons and spearheaded by a few armoured and mechanized units racing ahead. In order to maximize the capacity of its meagre motor transport, the organic transport of the armoured spearheads actually backtracked over the route of advance to pick up containerized fuel from prepositioned dumps—a novel modification of the staged resupply system. Motor transport was also supplemented by use of captured Soviet railroads (which had to be converted from wide to narrow gauge to accommodate German rolling stock) extending into the combat zone and paralleling vehicle roads.

The logistics of the North African desert campaigns in World War II virtually eliminated local supply and intermediate bases and depots, in effect replacing staged resupply by a simple single-shuttle base-to-troops operation. In 1941–42 the German Afrika Korps in Libya was supplied across the Mediterranean through the small port of Tripoli and eastward over a single coastal road that had no bases or magazines and was exposed to enemy air attack—a distance of up to 1,300 miles (2,100 km), depending on the location of the front (200 miles, or 320 km, was considered the normal limit for effective supply). This operation was occasionally supplemented by small coastal shipments into the ports of Banghāzī and Tobruk. The fuel cost of this overland operation was between one-third and one-half of all the fuel imported.

One of the striking lessons of World War II, often obscured by the tactical achievements of air power and mechanized armour, was the great power that modern logistics gave to the defense. In 1943 and 1944 the ratio of superiority enjoyed by Germany's enemies in output of combat munitions was about 2.5:1; the whole apparatus of Germany's war economy was subjected to relentless attack from the air and had to make good enormous losses of matériel in a succession of military defeats. Yet Germany was able, for about two years, to hold its own, primarily because its waning logistic strength could be concentrated on sustaining the firepower of forces that were stationary or retiring slowly toward their bases, instead of on the expensive effort required to support a rapid forward movement.

Logistic Specialization

For many centuries the soldier was a fighting man and nothing else; he depended on civilians to provide the services that enabled him to live, move, and fight. Even the more technical combat and combat-related skills, such as fortification, siegecraft, and service of artillery, were traditionally civilian. After the mid-19th century, with the rather sudden growth in the technical complexity of warfare, the military profession faced the problem of assimilating a growing number and variety of noncombatant skills. Many of the uniformed logistic services date from this period; examples are the British army's Transport Corps (later the Royal Army Service Corps), Hospital Corps, and Ordnance Corps. In the American Civil War the Union army formed a railway construction corps, largely civilian but under military control. A little later, Prussia created a railway section in the Great General Staff and a combined military–civilian organization for controlling and operating the railroads in time of war.

Not until the 20th century, however, did organized military units performing specialized logistic services begin to appear in large numbers in the field. By the end of World War II, what was called "service support" comprised about 45 percent of the total strength of the U.S. Army. Only three out of every 10 soldiers had combat functions, and even within a combat division one man out of four was a noncombatant. Even so, the specialized services that the military profession

MULBERRY

Mulberry was a code name used for either of two artificial harbours designed and constructed by the British in World War II to facilitate the unloading of supply ships off the coast of Normandy, France, immediately following the invasion of Europe on D-Day, June 6, 1944. One harbour, known as Mulberry A, was constructed off Saint-Laurent at Omaha Beach in the American sector, and the other, Mulberry B, was built off Arromanches at Gold Beach in the British sector. Each harbour, when fully operational, had the capacity to move 7,000 tons of vehicles and supplies per day from ship to shore.

Each Mulberry harbour consisted of roughly 6 miles (10 km) of flexible steel roadways (code-named Whales) that floated on steel or concrete pontoons (called Beetles). The roadways terminated at great pierheads, called Spuds, that were jacked up and down on legs that rested on the seafloor. These structures were to be sheltered from the sea by lines of massive sunken caissons (called Phoenixes), lines of scuttled ships (called Gooseberries), and a line of floating breakwaters (called Bombardons). It was estimated that construction of the caissons alone required 330,000 cubic yards of concrete, 31,000 tons of steel, and 1.5 million yards of steel shuttering.

One of the floating Mulberry roadways being put to good use at Omaha Beach in Normandy, France. Keystone/Hulton Archive/Getty Images

The Mulberry harbours were conceived after the failed amphibious raid on the French port of Dieppe in August 1942. The German defense of the coast of western Europe was built on formidable defenses around ports and port facilities. Because of the strength of these defenses, the Allies had to consider other means to push large quantities of provisions across the beaches in the early stages of an invasion. The British solution to the problem was to bring their own port with them. This solution had the support of Prime Minister Winston Churchill, who in May 1943 wrote the following note:

> *Piers for use on beaches: They must float up and down with the tide. The anchor problem must be mastered.... Let me have the best solution worked out. Don't argue the matter. The difficulties will argue for themselves.*

With Churchill's support, the artificial harbours received immediate attention, resources, time, and energy.

The various parts of the Mulberries were fabricated in secrecy in Britain and floated into position immediately after D-Day. Within 12 days of the landing (D-Day plus 12), both harbours were operational. They were intended to provide the primary means for the movement of goods from ship to shore until the port at Cherbourg was captured and opened. However, on June 19 a violent storm began, and by June 22 the American harbour was destroyed. (Parts of the wreckage were used to repair the British harbour.) The Americans had to return to the old way of doing things: bringing landing ships in to shore, grounding them, off-loading the ships, and then refloating them on the next high tide. The British Mulberry supported the Allied armies for 10 months. Two and a half million men, a half million vehicles, and four million tons of supplies landed in Europe through the artificial harbour at Arromanches. Remains of the structure can be seen to this day near the Musée du Débarquement.

succeeded in assimilating were only a small fraction of those on which the combat soldier depended. Throughout the vast administrative zones behind combat areas and in the national base, armies of civilian workers and specialists manned depots, arsenals, factories, communication centres, ports, and the other apparatuses of a modern society at war. Military establishments employed growing numbers of civilian administrators, scientists, technicians, management and public relations experts, and other specialists. Within the profession itself, the actual incorporation of specialized skills was limited, in the main, to those directly related (or exposed) to combat, such as the operating and servicing of military equipment, though even there the profession had no monopoly. Soldiers also served as administrators and supervisors over civilian specialists with whose skills they had only a nodding acquaintance. On the whole, the fighting man at mid-20th century belonged to a shrinking minority in a profession made up largely of administrators and noncombatant specialists.

LOGISTICS IN THE NUCLEAR AGE

The dropping of the first atomic bombs in August 1945 seemed to inaugurate a new era in warfare, demanding radical changes in logistic systems and techniques. The bombs did, in truth, give birth to a new line of weaponry of unprecedented destructive power. Within a decade they were followed by the thermonuclear weapon, an even greater leap in destructive force. Development of intercontinental ballistic missiles and nuclear-powered, missile-firing submarines a few years later extended the potential range of destruction to targets anywhere on the globe. The following decades saw dramatic developments in the offensive capabilities of nuclear weapons and also, for the first time, in defenses against them. But the world moved into the late 20th century without any of the new nuclear weaponry having been used in anger. Most warfare, moreover, was limited in scale and made little use of advanced technology. It produced only nine highly mobilized war economies: the two Koreas (1950–53), Israel (1956, 1967, 1973), North Vietnam (1965–75), Biafra (1967–70), Iran and Iraq (1980–88)—all except Israel preindustrial Third World countries.

The first major conflict in this period, the war in Korea (1950–53), seemed in many ways an extension of the positional campaigns in World War II. It was fought largely with World War II weapons, in some cases improved versions, and with stocks of munitions left over from that conflict. United Nations forces had an excellent base in nearby Japan, whose factories made a major contribution by rebuilding U.S. World War II material. UN air superiority kept both Japan and Pusan, South Korea's major port of entry, free from communist air attack. UN forces thus were able to funnel through Pusan supply tonnages comparable to those handled by the largest ports in World War II and to concentrate depots and other installations in the Pusan area to a degree that would have been suicidal without air superiority. The communist supply system, although technically primitive, functioned well under UN air attack, moving troops and supplies by night, organizing local labour, and exploiting the Chinese soldier's famous ability to fight well under extreme privation.

By World War II standards, the Korean War was a limited conflict (except for the two Korean belligerents, on whose soil it was fought). It involved only a partial, or "creeping," economic mobilization in the United States and a modest mobilization of reserves. Yet this was no small war. Over three years about 37.2 million measurement tons of cargo were poured into the South Korean ports, more than three-fourths of the amount shipped to U.S. Army forces in all the Pacific theatres in World War II. Combined UN forces reached a peak strength of almost one million men; communist forces were considerably larger.

NEW TECHNOLOGY

Advances in the technology of supply and movement after 1945 were not commensurate with those in weaponry. On land, internal-combustion vehicles and railroads, with increasing use of diesel fuel in both, remained the basic instruments of large-scale troop and freight movement despite their growing vulnerability to attack. In the most modern systems, substantial amounts of motor transport were capable of crossing shallow water obstacles. In areas not yet penetrated by rail or metaled roads—areas where much of the warfare of the period occurred—surface movement necessarily reverted to the ancient modes of human and animal porterage, sometimes usefully supplemented by the bicycle. Some exotic types of vehicles capable of negotiating rough and soft terrain off the roads were designed and tested—the "hovercraft," or air-cushion vehicle, for instance. But none of these innovations came into general use. The most promising developments in overland movement were helicopters and vertical-takeoff-and-landing aircraft, along with techniques of rapid airfield construction, which enabled streamlined airmobile forces and their logistic tails to overleap terrain obstacles and greatly reduced their dependence on roads, airfields, and forward bases. Helicopters also permitted the establishment and maintenance of isolated artillery fire bases in enemy territory.

In air movement there was a spectacular growth in the range and payload capacity of transport aircraft. The piston-engine transports of World War II vintage that carried out the Berlin airlift of 1948–49 had a capacity of about four tons and a maximum range of 1,500 miles (2,400 km). The U.S. C-141 jet transport, which went into service in 1965, had a 45-ton capacity and a range of 3,000 miles (4,800 km); it could take an average payload of 24 tons from the U.S. West Coast to South Vietnam in 43 hours and evacuate wounded back to the East Coast (10,000 miles, or 16,000 km) in less than a day. By 1970 these capabilities were dwarfed by the new "global logistics" C-5A, with payloads up to 130 tons and ranges up to 5,500 miles (8,800 km). It is estimated that 10 C-5As could have handled the entire Berlin airlift, which employed more than 140 of the then-available aircraft. C-5As played a vital role in the U.S. airlift to Israel during the Arab-Israeli War of October 1973. Very large cargo helicopters were also developed, notably in the Soviet Union, as were new techniques for packaging and air-dropping cargo.

In this period, movement by sea was the only branch of logistics that tapped the huge potential of nuclear propulsion. Its principal application, however, was in submarines, which did not develop a significant logistic function. (Development of nuclear-powered aircraft proved abortive.) The Soviet Union produced a nuclear-powered icebreaker in 1957, and the United States launched the first nuclear-powered merchant ship in 1959. But high initial and operating costs and (in the West) vested mercantile interests barred extensive construction of nuclear merchant ships. Except for supertankers built after the Suez crisis in 1956, and again during the energy crisis of the 1970s, seaborne cargo movement still depended on ships not radically different from those used in World War II. The chief technical improvement in sea lift, embodied in a few special-purpose vessels, was the "roll-on-roll-off" feature, first used in World War II landing craft, which permitted loading and discharge of vehicles without hoisting. Containerization, the stowage of irregularly shaped freight in sealed, reusable containers of uniform size and shape, became widespread in commercial ship operations and significantly affected ship design.

This period saw further development, from World War II models, of large vessels capable of discharging landing craft and vehicles offshore or over a beach as well as transporting troops, cargo, and helicopters in amphibious operations. For follow-up operations, improved attack cargo ships were built, such as the British landing-ship logistic, with accommodations for landing craft, helicopters, vehicles and tanks, landing ramps, and heavy-cargo-handling equipment. More revolutionary additions to the technology of amphibious logistics were the American landing vehicle hydrofoil and the BARC, both amphibians with pneumatic-tired wheels for overland movement and, in the latter case, capacity for 100 tons of cargo. Hydrofoil craft, which skimmed at high speeds above the water on submerged inclined planes, developed a varied family of types by 1970.

The revolution in electronic communication after World War II also had a profound impact on logistic administration. In advanced logistic systems the combination of advanced electronic communication with the high-speed electronic computer almost wholly replaced the elaborate processes of message transmission, record search, and record keeping formerly involved in supply administration, making the response of supply to demand automatic and virtually instantaneous.

Strategic Mobility

Because the leading military powers did not directly fight each other during the decades after World War II, none of them had to deal with the classic logistic problem of deploying and supporting forces over sea lines of communication exposed to enemy attack. The Soviet Union was able in 1962 to establish a missile base in Cuba manned by some 25,000 troops without interference by the United States until its offensive purpose was detected. Similarly, the large deployments of U.S. forces to Korea, Southeast Asia, and elsewhere, as well as the 8,000-mile (12,800-km) movement of a British expeditionary force to the Falkland Islands in 1982, encountered no opposition.

Yet the problem of strategic mobility was of major concern after 1945 to the handful of nations with far-flung interests and the capacity to project military power far beyond their borders. In the tightly

Diego Garcia

When the United States and its allies launched their attack on Iraq in 2003, Diego Garcia—home to U.S. long-range bombers, patrol planes, and cargo ships as well as refueling and other support personnel—once again proved its logistic value as what many considered to be one of the top three U.S. military bases in the world. The Persian Gulf War of 1990–91 and the war in Afghanistan, launched in 2001, had already demonstrated the military importance of the Indian Ocean atoll as a naval and air force base and observatory (both satellite and communications). From a geostrategic point of view, the Diego Garcia atoll, located in the Chagos Archipelago, or British Indian Ocean Territory (BIOT), boasts undeniable advantages—a lagoon of considerable size and depth; a natural port able to accommodate ships, aircraft carriers, and both conventional and nuclear submarines; and an ideal location in a cyclone-free zone in close proximity to international shipping lanes. Such advantages made a military stronghold of Diego Garcia, and during the 1970s and '80s it became the largest British-American naval support base in the Indian Ocean.

The United Kingdom had bought the Chagos Archipelago in November 1965 from Mauritius, then a British crown colony. The deal was accepted without much negotiation by Mauritius Chief Minister Sir Seewoosagur Ramgoolam, whose primary objective was to achieve independence (obtained in 1968). The strategically placed BIOT initially comprised the Chagos Archipelago and three other islands belonging to the Seychelles. After the Seychelles gained independence in 1976–77, London returned the three islands.

The militarization of Diego Garcia was the result of three successive bilateral treaties between the United Kingdom and the United States between 1966 and 1976. In the treaty of Dec. 12, 1966, control of Diego Garcia was handed over to the U.S. for 50 years, renegotiable for an additional 20 years. Although the British retained sovereignty with an on-site flagship, the administration was American. Thanks to the treaty of Feb. 25, 1976, a naval support base was officially installed, which allowed the U.S. Navy a permanent outpost in the Indian Ocean. This new development elicited protests, notably from the Soviet Union; the UN, which in 1971 had approved a resolution by Sri Lanka and India declaring the Indian Ocean a "peace zone"; and Mauritius, which initiated an annual debate in parliamentary hearings and international forums regarding the retrocession of the Chagos Islands.

Between 1967 and 1973 some 1,400 (estimates varied) Chagos islanders, called Ilois, were expelled to live in Mauritius and Seychelles. In 1976 the U.S. ordered the systematic displacement of the remaining local people on Diego Garcia and replaced them with a temporary staff brought in from Mauritius and Seychelles.

controlled power politics of the period, each of these countries needed the capability to bring military force quickly to bear to protect its interests in local emergencies at remote points—as Great Britain and France did at Suez, Egypt, in 1956, the United States in Lebanon in 1958 and in the Taiwan Straits in 1959, Great Britain in Kuwait in 1961 and in the Falkland Islands in 1982, and France in Chad on several occasions in the 1980s. The most effective instruments for such interventions were small, powerful, mobile task forces brought in by air or sea as well as forward-deployed aircraft-carrier and amphibious forces. The United States developed strong and versatile intervention capabilities, with major fleets deployed in the far Pacific and the Mediterranean; a worldwide network of bases and alliances; large ground and air forces in Europe, Korea, and Southeast Asia; and, in the 1960s, a mobile strategic reserve of several divisions with long-range sea-lift and airlift capabilities. The Soviet Union, Great Britain, and France had more limited capabilities, although the Soviet Union began in the late 1960s to deploy strong naval and air forces into the eastern Mediterranean and also maintained a naval presence in the Indian Ocean. After the U.S. withdrawal from Vietnam in 1973, the Soviet navy extended its power into the South China Sea.

The logistics of strategic mobility was complex and was decisively affected by the changing technology of movement, especially by air and sea. During the 1950s the proponents of naval and land-based air power debated the relative cost and effectiveness of naval-carrier forces and fixed air bases as a tool of emergency intervention. Studies seemed to show that the fixed bases were cheaper if all related costs were considered but that the advantage of mobility and flexibility lay with the naval carriers. In the 1970s the growing range and capacities of transport aircraft provided an increasingly effective tool for distant intervention and were a large factor in the reduction of the American and British overseas base systems. In practice, emergency situations called for using the means available and involved a great deal of improvisation, especially for second-rank powers.

MANAGEMENT

Both during and after World War II the United States operated the largest and most advanced logistic system in the world. Its wartime operations stressed speed, volume, and risk-taking more than efficiency and economy. The postwar years, with accelerated technological change, skyrocketing costs, and diminished public interest in defense, brought a revulsion against military prodigality, manifested by calls for reduced defense budgets and a growing demand for more efficient management of the military establishment. This demand culminated in a thorough overhaul of the whole system in the 1960s.

One result was the reorganization of logistic activities in the three military services, generally along functional lines, with large logistic commands operating under functional staff supervision. In each service, however, each major weapon system was centrally managed by a separate project officer, and central inventory control was maintained for large commodity groups. In 1961 a new defense supply agency was established to manage on a wholesale basis the procurement, storage, and distribution of common military supplies and the administration of certain common services.

The most far-reaching managerial reforms of the period were instituted by the U.S. defense secretary, Robert S. McNamara (1961–68), in the resource allocation process. A unified defense planning–programming–budgeting system provided for five-year projections of force, manpower, and dollar requirements for all defense activities, classified into eight or nine major programs (such as strategic forces) that cut across the lines of traditional service responsibilities. The system was introduced in other federal departments after 1965, and elements of it were adopted by the British and other governments. In 1966 a program was inaugurated to integrate management accounting at the operating level with the programming–budgeting system. At the end of the 1960s a new administration restored some of the initiative in the planning–budgeting–programming cycle to the Joint Chiefs of Staff and the military services.

The reforms of the 1960s exploited the whole range of current managerial methodology. The basic techniques, such as systems and operations analysis, all stressed precise, scientific, usually quantitative formulations of problems and mathematical approaches to rational decision making. Systems analysis, the technique associated with defense planning and programming, was a method of economic and mathematical analysis useful in dealing with complex problems of choice under conditions of uncertainty. The technological foundation of this improved logistic management was the high-speed electronic computer, which was being used chiefly in inventory control; in automated operations at depots, bases, and stations; in transmitting and processing supply data; in personnel administration; and in command-and-control networks.

War in Vietnam

One of the most significant developments in logistics after 1945 was the pitting of advanced high-technology systems against well-organized low-technology systems operating on their own ground. The Korean War and the anticolonial wars in French Indochina and Algeria were the principal conflicts of this kind in the 1950s. The war in Vietnam following large-scale U.S. intervention in 1965 brought into conflict the most effective of both types of systems.

Because South Vietnam lacked most of the facilities on which modern military forces depend, the massive U.S. deployment that began in the spring of 1965, reaching 180,000 men by the end of

Viet Cong soldier standing with an AK-47, February 1973. SSGT Herman Kokojan/Department of Defense Media (DD-ST-99-04298)

that year and more than 550,000 in 1969, was accompanied, rather than preceded, by a huge ($4 billion) construction program, carried out partly by army, navy, and air force engineer units and partly by a consortium of engineering contractors. Under this program were built seven deepwater and several smaller ports, eight jet air bases with 10,000-foot (3,050-metre) runways, 200 smaller airfields, and 200 heliports, besides millions of square feet of covered and refrigerated storage, hundreds of miles of roads, hundreds of bridges, oil pipelines and tanks, and all the other apparatuses of a modern logistic infrastructure. Deep-draft shipping brought in all but scarce items of airlifted supplies and came mainly from the U.S. directly.

The soldier in the field received lavish logistic support. By means of helicopter supply, troops in contact with the enemy were often provided with hot meals; most of the wounded were promptly evacuated to hospitals and serious cases were moved by air to base facilities in the Pacific or the United States. Medical evacuation, combined with advances in medicine, helped to raise the ratio of surviving wounded to dead to 6:1, in contrast to a World War II ratio of 2.6:1. Logistic support of army forces was organized under a single logistic command having a strength of 30,000 and employing 50,000 Vietnamese, U.S., and foreign civilians. Ultimately there were four or five support personnel for every infantryman who bore the brunt of contact fighting with the enemy.

The communist logistic system centred in the highly mobilized society of North Vietnam. In its integration, efficiency, and resilience under concentrated and prolonged bombing it rivaled the war economy of Germany in World War II. Its resilience owed much, however, to its being a village-centred agricultural society, with modest material needs and a limited industrial base, which produced no steel, very little pig iron, and only one-fifth as much electric power as a single power plant in a small American town.

By late 1967 the communist war effort in South Vietnam depended heavily on the flow of troops, equipment, and supplies from North Vietnam, supplied mainly by the Soviet Union. The troops and most of the supplies moved over the Ho Chi Minh Trail, originally a network of footpaths and dirt roads (often paved after 1967) through communist-controlled areas in Laos and Cambodia. Supplies also came into South Vietnam by sea, directly across the northern border, and, especially after 1967, through the Cambodian port of Kompong Som and overland into the Mekong delta.

The Ho Chi Minh Trail was a long, slow-moving pipeline, requiring from three to six months in transit by truck, barge, ox cart, bicycle, and foot, but its capacity was ample for the modest demands placed upon it. In mid-1967, U.S. intelligence estimated the total nonfood requirements of all communist forces in South Vietnam, except in the northernmost provinces, to be as low as 15 tons per day (about 1.5 ounces, or 43 grams, per man); food was procured locally and in nearby Cambodia

Mujahideen standing beside the debris of a helicopter they shot down with a Stinger missile in Maidan Province in Afghanistan in June 1987. AFP/Getty Images

and Laos. In 1968, when the pace of the war quickened and communist forces were substantially augmented, estimated non-food requirements rose to about 120 tons per day. (A single U.S. division required about five times this amount.)

American bombing had little effect on the flow of troops to the south, and the communist logistic system stood up remarkably well—and ultimately victoriously—under the weight of American air power. Its strength lay primarily in its austerity, but also in efficient organization, lavish use of manpower, availability of sanctuary areas in Laos and Cambodia, and a steady flow of imported supplies.

The Soviets in Afghanistan

The Soviet Union's Afghan war (1979–89), though on a scale smaller than Vietnam, embodied similar political, social, and economic dynamics and a similar contest between high-technology and low-technology logistic systems. Soviet forces, concentrated in the principal cities and towns, relied heavily on airlift and convoyed motor transport to move troops and supplies. Afghan guerrillas (called mujahideen), holding most of the countryside, used mainly animal transport and brought much of their supplies and weapons across the border from Pakistan. In an agriculturally poor country, significantly depopulated by Soviet bombing and forced flight into Pakistan, mass hunger and disease were widespread. For most of the war an approximate stalemate prevailed, in logistics as well as in tactical operations. But in 1986 the acquisition from the United States and Great Britain of substantial numbers of shoulder-fired surface-to-air missiles enabled the mujahideen to challenge Soviet control of the air—a significant factor in the Soviets' withdrawal early in 1989.

Growing Complexity

For logisticians the fundamental dilemma posed by the quantum leap in weapons technology after World War II was the absence of any comparable development in logistics. The electronic computer had, indeed, a dramatic impact on logistic planning and administration, as well as on military administration in general. The computer enabled planners to visualize problems concretely, often in quantitative terms; it accelerated the transmission of demand and the administrative response to it; and it enabled the military services for the first time to control their inventories. But the computer could not touch the ancient problem, compounded by the new weaponry, of actually providing and moving supplies to their users.

Conversely, nuclear weapons threatened to sweep away every vestige of the logistic system of the industrial era. None of the elaborate apparatuses of rear-area administration, lines of communication, or even sources of supply seemed likely to survive the nuclear firepower that could be brought to bear against it. The problem was studied and restudied, and a great deal of hopeful doctrine was developed for logistic operations in a nuclear war. It revolved about such concepts as dispersion, mobility, small targets, duplication, multiplicity, austerity, concealment, and automaticity, yet all of it was little more than a planner's dream, and a fading dream at that. At best it promised to reduce somewhat the inherent vulnerability of the surface-bound installations and transport on which military forces for the foreseeable future were likely to depend. Dispersion and duplication were enemies of economy and efficiency. The net effect could only be to increase the costs of logistic support and diminish the yield of delivered supplies and services.

In any case, nuclear war seemed the least likely of prospects. The most likely appeared to be a continuation of the confused patterns of limited conventional war and quasi-war that had filled the decades since the end of World War II. Under these conditions the central problems of logistics would be the historic ones of weight and bulk, which inhibited mobility and range of movement and were the primary causes of vulnerability to the new firepower. The technologies of these decades had accelerated the basic logistic trends of the industrial era: increasing complexity and cost in military hardware, increasing overall weight and volume of material (despite a reverse trend toward reduced numbers in some major items, such as aircraft), and, above all, an enormous increase in expenditures of ammunition and fuel. Logisticians in the postwar decades had to face the probability that in another large-scale conventional conflict between advanced powers the new vehicles would consume about half again as much fuel and the new weapons would expend more than four times as much ammunition as had been consumed and expended in World War II.

Some of the new tools of logistics were highly effective in specialized environments, notably the growing family of helicopters used in conjunction with conventional and short-takeoff-and-landing air transports, which permitted a mobility and a range of movement over difficult terrain far beyond the capabilities of surface transport. Whether an airmobile logistic system could survive the firepower likely to be encountered in a conflict with an adversary disputing command of the air was a question to which experience had not yet given an answer. In any case, the system purchased its mobility and range at a fuel cost several times higher than that involved in surface transport.

How well the "sophisticated" systems, with their growing burden of weight and bulk, would function under a threat to their previously immune supply lines was perhaps the most serious challenge facing modern logisticians. Nuclear propulsion offered a theoretical solution, but there seemed little hope for its early application to large sectors of military movement. A nuclear-powered sea transport service was a reasonable prospect, though not an early one, and it would not suffice for a major overseas war. More fundamentally, fuel consumption on the sea lanes was not the crux of the problem, and nuclear propulsion offered no solution to the vulnerability of surface vessels to air and submarine attack. The massive fuel consumers were aircraft and ground vehicles, and serious technical obstacles barred the application of nuclear energy to their power plants.

The reckoning, if there was to be one, might be long postponed. Given the existing distribution and equilibriums of power among the advanced nations, on the one hand, and the high cost and slow diffusion of sophisticated military technology to the less-developed two-thirds of the world, on the other, limited warfare seemed likely for a long time to come to remain at relatively low technical levels. Meanwhile, sophisticated logistic systems were becoming more entangled in their own complexity and absorbed in the endless pursuit of efficient management and in the struggle to control the waste and friction involved in delivering the tools of war to their users.

CHAPTER 4
GUERRILLA WARFARE

Guerrilla warfare is a type of warfare fought by irregulars in fast-moving, small-scale actions against orthodox military and police forces and, on occasion, against rival insurgent forces, either independently or in conjunction with a larger political-military strategy. The word *guerrilla* (the diminutive of the Spanish *guerra*, "war") stems from the duke of Wellington's campaigns during the Peninsular War (1808–14), in which Spanish and Portuguese irregulars, or *guerrilleros*, helped drive the French from the Iberian Peninsula. Over the centuries the practitioners of guerrilla warfare have been called rebels, irregulars, insurgents, partisans, and mercenaries. Frustrated military commanders have consistently damned them as barbarians, savages, terrorists, brigands, outlaws, and bandits.

The French military writer Henri, baron de Jomini, classified the operations of guerrilla fighters as "national war." The Prussian theorist Carl von Clausewitz reluctantly admitted their existence by picturing partisans as "a kind of nebulous vapoury essence." Later writers called their operations "small wars." During the Cold War (1945–91), Chinese leader Mao Zedong's term *revolutionary warfare* became a staple, as did *insurgency, rebellion, insurrection, people's war,* and *war of national liberation.*

HISTORY

Regardless of terminology, the importance of guerrilla warfare has varied considerably throughout history. Traditionally, it has been a weapon of protest employed to rectify real or imagined wrongs levied on a people either by a ruling government or by a foreign invader. As such, it has scored remarkable successes and has suffered disastrous defeats.

The role of guerrilla warfare considerably expanded during World War II, when Josip Broz Tito's communist Partisans tied down and frequently clashed with the German army in Yugoslavia and when other groups, both communist and noncommunist, fought against the German and Japanese enemies. During the prolonged Cold War period, numerous guerrilla forces of varying political beliefs were showered with money, modern weapons, and equipment from assorted benefactors. The stew of animosities was further seasoned by ethnic and religious rivalries, a factor that helps to explain why guerrilla warfare continues to be fought in a large number of countries today. In some instances it has assumed a universal character under the banner of religious fundamentalism. The most prominent practitioner of this type is the Muslim group al-Qaeda, which has attracted religious fanatics from various countries to carry out vicious terrorist attacks, the most famous being the September 11 attacks on the United States in 2001. Still another major change has been the transition of some guerrilla groups, notably in Colombia, Peru, Northern Ireland, and Spain, into criminal terrorism on behalf of drug barons and other Mafia-style overlords.

"Barbarian Archer in Scythian Costume," Athenian plate by Epictetus, late 6th century BCE; in the British Museum, London. In the 1st millennium BCE the Scythians, rulers of lands in Central Asia and north of the Black Sea, were admired and feared for their prowess with the bow. Courtesy of the trustees of the British Museum

Early History

In 512 BCE the Persian warrior-king Darius I, who ruled the largest empire and commanded the best army in the world, bowed to the hit-and-run tactics of the nomadic Scythians and left them to their lands beyond the Danube. Alexander the Great (356–323 BCE) also fought serious guerrilla opposition, which he overcame by modifying his

Mongol warriors, miniature from Rashīd al-Dīn's History of the World, *1307; in the Edinburgh University Library, Scotland.* Courtesy of the Edinburgh University Library, Scotland

tactics and by winning important tribes to his side. The Romans fought against guerrillas in their conquest of Spain for more than 200 years before Spain was truly subdued.

Guerrilla and quasi-guerrilla operations were employed in an aggressive role in ensuing centuries by such predatory barbarians as the Goths and the Huns, who forced the Roman Empire onto the defensive; the Magyars, who conquered Hungary; the hordes of northern barbarians who attacked the Byzantine Empire for more than 500 years; the Vikings, who overran Ireland, England, and France; and the Mongols, who conquered China and terrified central Europe. In the 12th century the Crusader invasion of Syria was at times stymied by the guerrilla tactics of the Seljuq Turks, a frustration shared by the Normans in their conquest of Ireland (1169–75). A century later, Kublai Khan's army of Mongols was driven from the area of Vietnam by Tran Hung Dao, who had trained his army to fight guerrilla warfare. King Edward I of England struggled through long, hard, and expensive campaigns to subdue Welsh guerrillas; that he failed to conquer Scotland was largely due to the brilliant guerrilla operations of Robert the Bruce (Robert I; 1306–29). Bertrand du Guesclin, a Breton guerrilla leader in the Hundred Years' War (1337–1453), all but pushed the English from France by using Fabian (cautious and slow) tactics of harassment, surprise, ambush, sudden assault, and slow siege.

Origins of Modern Guerrilla Warfare

Guerrilla warfare in time became a useful adjunct to larger political and military strategies—a role in which it complemented orthodox military operations both inside enemy territory and in areas seized and occupied by an enemy. Early examples of this role occurred in the first two Silesian Wars (1740–45) and in the Seven Years' War (1756–63), when Hungarian, Croatian, and Serbian irregulars (called *Grenzerer*, "border people"), fighting in conjunction with the Austrian army, several times forced Frederick the Great (Frederick II) of Prussia to retreat from Bohemia and Moravia after suffering heavy losses. Toward the end of the U.S. War of Independence (1775–83), a ragtag band of South Carolina irregulars under Francis Marion relied heavily on terrorist

Zaporozhian Cossacks, *oil painting by Ilya Yefimovich Repin, 1891; in the State Russian Museum, St. Petersburg. Repin's famous historical painting re-creates the drafting of a mocking and insulting letter in 1679 to Ottoman sultan Mehmed IV, who had demanded a Cossack surrender.* Novosti Press Agency

tactics to drive the British general Lord Cornwallis from the Carolinas to defeat at Yorktown, Virginia. Wellington's operations in Spain were frequently supported by effectively commanded regional bands of guerrillas—perhaps 30,000 in all—who made life miserable for the French invaders by blocking roads, intercepting couriers, and at times even waging conventional war. In 1812, in the long retreat from Moscow, the armies of Napoleon I suffered thousands of casualties inflicted by bands of Russian peasants working with mounted Cossacks.

Guerrilla wars flourished in the following two centuries as native irregulars in India, Algeria, Morocco, Burma (Myanmar), New Zealand, and the Balkans tried, usually in vain, to prevent colonization by the great powers. Indian tribes in North America stubbornly fought the opening of the West; Cuban guerrillas fought the Spanish; and Filipino guerrillas fought the Spanish and Americans. In the South African War 90,000 Boer commandos held off a large British army for two years before succumbing.

Comanche Mounted War Party, *oil on canvas by George Catlin, 1834; in the National Museum of American Art, Smithsonian Institution, Washington, D.C.* Courtesy of the Smithsonian American Art Museum (formerly National Museum of American Art), Washington, D.C., gift of Mrs. Sarah Harrison

Pancho Villa on horseback, 1916. Encyclopædia Britannica, Inc.

As these bloody campaigns continued, political motivations became more and more important. The Taiping Rebellion (1850–64) in China, a peasant uprising against the Qing dynasty, killed an estimated 20 million Chinese before it was suppressed. During the American Civil War mounted guerrillas from both sides raided far behind enemy lines, often looting and pillaging randomly. Mexican peasants, fighting under such leaders as Emiliano Zapata and Pancho Villa, used guerrilla warfare to achieve a specific political goal in the Mexican Revolution (1910–20). Arab tribesmen under Fayṣal I employed the brilliant guerrilla strategies and tactics of British officer T.E. Lawrence in their campaign to liberate their lands from the Ottoman Empire in World War I. In 1916 the Easter Rising in Ireland led to a ferocious guerrilla war fought by the Irish Republican Army (IRA)—a war that ceased only with the uneasy peace and partition of Ireland in 1921. In 1927 communist leader Mao Zedong raised the flag of a rural rebellion that continued for 22 years. This experience resulted in a codified theory of protracted revolutionary war, Mao's *On Guerrilla Warfare* (1937), which was later called "the most radical, violent and extensive theory of war ever put into effect."

The Cold War Period

Political ideology became a more pronounced factor in the numerous guerrilla campaigns of World War II. In most of the countries invaded by Germany, Italy, and Japan, local communists either formed their own guerrilla bands or joined other bands—such as the French and Belgian resistance fighters who called their organization *Le Maquis* (meaning "underbrush"). While consolidating their hold on the country, some of these groups spent as much time eliminating indigenous opposition as they did fighting the enemy, but most of them contributed sufficiently to the Allied war effort to be sent shipments of arms, equipment, and gold, which helped them to challenge existing governments after the war. In the following decades the Soviet Union and United States supported a series of widespread guerrilla insurgencies and counterinsurgencies in dangerous and often unproductive—but always costly—proxy wars.

In Yugoslavia and Albania the communist takeover of government was simple and immediate; in China it was complicated and delayed; in South Vietnam it succeeded after nearly three decades; in Greece, Malaysia, and the Philippines it was foiled—but only after prolonged and costly fighting. Noncommunist insurgents simultaneously used guerrilla warfare, with heavy emphasis on terrorist tactics, to help end British rule in Palestine in 1948 and Dutch rule in Indonesia in 1949.

After 1948 the new state of Israel was faced with a guerrilla war conducted by the fedayeen (a term used in Islamic cultures to describe a devotee of a religious or national group willing to engage in self-immolation to attain a group goal) of its Arab neighbours—a protracted and vicious struggle that over the next 30 years led to three quasi-conventional wars (each an Israeli victory) followed by renewed guerrilla war. Despite concerted efforts to negotiate a peace, the struggle continued, as the Palestine Liberation Organization (PLO), its militant wing Fatah, and three competing major terrorist groups (Ḥamās, Islamic Jihad, and al-Aqṣā Martyrs Brigade) remained determined to regain control of the West Bank and Gaza Strip and, eventually (a long-term goal for at least some of them), all of pre-1948 Palestine.

Mao's victory in China in 1949 established him as the prophet of "revolutionary warfare" who had transferred Marxism-Leninism from the industrial areas to the countryside and in doing so heartened contemporary insurgents and encouraged new ones. In Indochina, Ho Chi Minh's Viet Minh guerrillas, ably commanded by Vo Nguyen Giap, had been fighting the French overlords since 1945. The struggle ended in 1954 with the Battle of Dien Bien Phu, when a strongly fortified French garrison surrendered after a two-month-long quasi-conventional ground attack by

GENERAL GIAP

Vo Nguyen Giap (born 1912, An Xa, Vietnam) was a Vietnamese military and political leader whose perfection of guerrilla as well as conventional strategy and tactics led to the Viet Minh victory over the French and later to the North Vietnamese victory over South Vietnam and the United States.

The son of an ardent anticolonialist scholar, Giap as a youth began to work for Vietnamese autonomy. He attended the same high school as Ho Chi Minh, the communist leader, and while still a student in 1926 he joined the Tan Viet Cach Menh Dang, the Revolutionary Party of Young Vietnam. In 1930, as a supporter of student strikes, he was arrested by the French Sûreté and sentenced to three years in prison, but he was paroled after serving only a few months. He studied at the Lycée Albert-Sarraut in Hanoi, where in 1937 he received a law degree. Giap then became a professor of history at the Lycée Thanh Long in Hanoi, where he converted many of his fellow teachers and students to his political views. In 1938 he married Minh Thai, and together they worked for the Indochinese Communist Party. When in 1939 the party was prohibited, Giap escaped to China, but his wife and sister-in-law were captured by the French police. His sister-in-law was guillotined; his wife received a life sentence and died in prison after three years.

In 1941 Giap formed an alliance with Chu Van Tan, guerrilla leader of the Tho, a minority tribal group of northeastern Vietnam. Giap hoped to build an army that would drive out the French and support the goals of the Viet Minh, Ho Chi Minh's Vietnamese independence movement. With Ho Chi Minh, Giap marched his forces into Hanoi in August 1945, and in September Ho announced the independence of Vietnam, with Giap in command of all police and internal security forces and commander in chief of the armed forces. Giap sanctioned the execution of many noncommunist nationalists, and he censored nationalist newspapers to conform with Communist Party directives. In the French Indochina War, Giap's brilliance as a military strategist and tactician led to his winning the decisive battle at Dien Bien Phu on May 7, 1954, which brought the French colonialist regime to an end.

On the division of the country in July, Giap became deputy prime minister, minister of defense, and commander in chief of the armed forces of North Vietnam. He subsequently led the military forces of the north to eventual victory in the Vietnam War, compelling the Americans to leave the country in 1973 and bringing about the fall of South Vietnam in 1975. From 1976, when the two Vietnams were reunited, to 1980 Giap served as Vietnam's minister of national defense; he also became a deputy prime minister in 1976. He was a full member of the Politburo of the Vietnamese Communist Party until 1982. Giap was the author of People's War, People's Army *(1961), a manual of guerrilla warfare based on his own experience.*

Giap's army. A civil war followed between Ho's North Vietnam and South Vietnam, the former supported by the Soviet Union and China and the latter by the United States and its allies. U.S. involvement in the Vietnam War steadily increased, resulting in the first commitment of U.S. troops in 1961 and ending only with the North Vietnamese conquest of the entire country in 1975.

Meanwhile, a spate of new insurgencies, both communist and noncommunist, followed to end French rule in Algeria and British rule in Kenya, Cyprus, and Rhodesia. Fidel Castro's overthrow of the tottering and corrupt regime of Fulgencio Batista in Cuba in 1959 provoked other rural insurgencies throughout Latin America, Asia, the Middle East, and Africa. Old and new insurgencies flourished in Peru, Colombia, El Salvador, Nicaragua, the Philippines, Thailand, Sri Lanka, India, Kashmir, Lebanon, Syria, Morocco, Angola, Mozambique, Northern Ireland, and Spain.

The Afghan War of 1978–92 saw a coalition of Muslim guerrillas known as the mujahideen, variously commanded by regional Afghan warlords heavily subsidized by the United States, fighting against Afghan and Soviet forces. The Soviets withdrew from that country in 1989, leaving the Afghan factions to fight it out in a civil war. South Africa similarly was forced to relinquish control of South West Africa (now Namibia) in 1989, and guerrilla activity by the African National Congress (ANC)—one of the most successful guerrilla operations of the modern era—was largely responsible for the end of the apartheid system and for the institution of universal suffrage in South Africa in 1994.

In the early 1970s the general failure of rural insurgencies in Central and South America caused some frustrated revolutionaries to shift from rural to urban guerrilla warfare with emphasis on the use of collective terrorism. Fired by the quasi-anarchistic teachings of German American political philosopher Herbert Marcuse, French revolutionary-philosopher Régis Debray, and others and armed with a do-it-yourself manual of murder (Carlos Marighela, *For the Liberation of Brazil* [1970]), New Left revolutionaries embraced assassination, robbery, indiscriminate bombing, and kidnapping to attain their ends—crimes that became the order of the day as did, on an international scale, airplane hijackings, kidnappings, and mass murder.

Such was the media-heightened impact of urban guerrilla warfare, and such its potential danger to civilized society, that some observers believed "urban terrorism" should be classified as a new genre of warfare. But terrorist tactics, urban or rural, even the most extreme, have always been integral to guerrilla and counterguerrilla warfare—indeed to all warfare. "Kill one, frighten 10,000," wrote the ancient Chinese general Sunzi.

Initially, urban guerrilla warfare alone appeared to be a losing proposition, in that its promiscuous collective destruction—particularly mass murder—tended to alienate a formerly passive and even sympathetic citizenry. Its Achilles' heel was threefold: a lack of a viable political goal based on the repair of social, economic, and political failures; a lack of an organization designed to reach that goal and capable of providing operational bases and sanctuary areas; and a failure to recruit and train new activists. The lack of organization in depth helps to explain the eventual demise of fringe advocates and practitioners of urban and international terrorism, groups far removed from guerrilla insurgencies. Examples of such groups in the 1970s and '80s are the Black Panther Party, the Weathermen, and the Symbionese Liberation Army in the United States; the Japanese Red Army; the Red Army Faction in West Germany; the Angry Brigade in the United Kingdom; the Red Brigades of Italy; Direct Action in France; and Middle Eastern groups such as the Popular Front for the Liberation of Palestine–General Command and the Abū Niḍāl Group.

However, urban warfare, once properly organized and combined with rural guerrilla warfare and with the increased employment of bomb attacks, played an important role in bringing cease-fires and even peace (however tentative) to such places as Northern Ireland, Sri Lanka, and Israel-Palestine (though not to Colombia, Spain, Indonesia, Nepal, the Philippines, or Chechnya). Not without reason did some experts conclude that guerrilla warfare and terrorism, rural or urban, internal or international, had become the primary form of conflict for that time.

THE POST-COLD WAR PERIOD

The collapse of the Soviet Union in 1991 did little to alter this gloomy prognostication. Variations of communist ideology, Marxist or Maoist, continued to fuel insurgencies in Colombia, Peru, Mexico, Spain, Sri Lanka, Turkey, Nepal, East Timor, and the Philippines. Added to this was the growth of the Muslim religious factor in such localized insurgencies as Israel-Palestine and Kashmir and in renegade terrorist organizations such as Osama bin Laden's al-Qaeda. Bin Laden, a wealthy Saudi Arabian expatriate and religious fanatic, patched together a worldwide network of followers whose activities during the 1990s and beyond included a series of hideous bombings. Forced to take refuge in Taliban-ruled Afghanistan, bin Laden planned the aerial suicide attacks of Sept. 11, 2001, on the United States. This deed led to the "war on terror."

Finally, on May 2, 2011, bin Laden was killed by U.S. forces in a raid on his compound in Abbottabad, Pakistan. The fate of al-Qaeda was thrown into question soon after the death of its leader.

PRINCIPLES

From the ancient Scythians to modern revolutionaries, guerrilla groups have survived by addressing certain fundamental principles: how to motivate their fighters, how to find shelter and support, choosing good leaders, building an effective organization, and finding the right weapons with which to attack the enemy.

Purpose and Motivation

Fundamental to militant revolution is a cause, which unfortunately has never been difficult to find in a less-than-perfect world. The guerrilla cause may assume several guises: to the world it may be presented as liberating a country from a colonial yoke or from an invader's rule; to the peasant it may be freedom from serfdom, from oppressive rents to absentee landlords, or from taxation; to a middle-class citizen it may be establishment or restoration of representative government as opposed to a military or totalitarian dictatorship.

Whether real or artificial, whether inspired by political ideology, religion, nationalism, or, more often, a genuine desire for a better life, this cause is fundamental in motivating people to armed action. Mao leaves no doubt of its importance:

> *Without a political goal, guerrilla warfare must fail, as it must if its political objectives do not coincide with the aspirations of the people and their sympathy, cooperation, and assistance cannot be gained.*

The lack of a viable political goal has often been the key factor in an insurgency's failure. It will continue to be so as long as an insurgency is tainted by extreme criminal actions. Some insurgent leaders recognize this basic fact in confining revolutionary activities to their traditional purposes.

Popular Support

Revolutionary writings have constantly stressed the guerrillas' affiliation with the people. Guerrillas spring from the people, who in turn support their spawn, not only by furnishing sons and daughters to the cause but also by furnishing money, food, shelter, refuge, transport, medical aid, and intelligence—support that must simultaneously be denied to the enemy. Although T.E. Lawrence called for no more than "a friendly population, not actively friendly, but sympathetic to the point of not betraying rebel movements to the enemy," he also wrote that his guerrillas "had won a province when the civilians in it had been taught to die for the ideal of freedom." Georgios Grivas, a Greek soldier who led the Cypriot rebellion in the 1950s, wrote that a guerrilla war stands no chance of success unless it has "the complete and unreserved support of the majority of the country's inhabitants." Mao repeatedly stressed the importance of proper troop behaviour: the Chinese guerrilla was required to pay a peasant for food, to respect his property, and not to offend propriety by undressing in front of a peasant woman.

Essential to maintaining domestic support and to gaining international support is vigorous, intelligent, and believable propaganda. Tito spread the word by newspaper and the Algerians by newspaper and radio, thereby enforcing Lawrence's dictum that the press is the greatest weapon in the army of a modern commander. The printed word has since been supplemented by the television camera, which has been defined as "a weapon lying in the street, which either side can pick up and use—and is more powerful than any other." Today images of guerrilla and counterguerrilla clashes are delivered in real time, via satellite television and the Internet, from around the world.

Leaders and Recruits

Such are the vicissitudes of guerrilla warfare that outstanding leadership is necessary at all levels if a guerrilla force is to survive and prosper. A leader must not only be endowed with intelligence and courage but must be buttressed by an almost fanatical belief in himself and his cause. Lawrence, Tito, Mao, Ho, Castro, the Soviet leaders Vladimir Ilich Lenin and Leon Trotsky, the Filipino Luis Taruc, the Israeli Menachem Begin, the Kenyan Jomo Kenyatta, the Malayan Ch'en P'ing, the Algerian Ahmed Ben Bella, the Palestinian Yāsir ʿArafāt, the Sri Lankan Vellupillai Prabhakaran, the East Timorese Xanana Gusmão, Osama bin Laden, a host of IRA leaders in Northern Ireland and ETA leaders in Spain—these and many others attracted followers to a cause, organized them, and instilled a disciplined zeal matched only by the most elite military organizations.

The guerrilla recruit must be resourceful and enduring, committed totally to the cause if he is to withstand the hardships and dangers of guerrilla fighting. A prolonged and difficult campaign may force guerrilla leaders to abandon selectivity and resort to intimidation in order to gain recruits—as was the case in Vietnam, where rigorous political indoctrination only partially compensated for lack of voluntary zeal.

Emiliano Zapata, the Agrarian Leader, *lithograph by Diego Rivera, 1932.* Library of Congress, Washington, D.C. (neg. no. LC-USZC4-3908)

Organization and Unity of Command

The tactical organization of guerrilla units varies according to size and operational demands. Mao called for a guerrilla squad of 9 to 11; his basic unit was the company, about 120 strong. Grivas initially deployed sabotage-terrorist teams of only four or five members. The Greek Civil War of the late 1940s opened with about 4,000 communist guerrillas divided into units of 150 fighters that, as strength increased, grew to battalions 250 strong. Tito began his campaign with about 15,000 fighters organized into small cadres; he ended the war with some 250,000 troops organized into brigades. Vietnamese guerrillas initially were organized into small squads that expanded to battalion and even regimental strengths. As modern guerrilla leaders have discovered, undue expansion may result in security failures and in partial loss of control, as has been the case in Northern Ireland, Colombia, and Palestine. Guerrilla units for the most part have remained small and more tightly organized in a cellular structure that, from a security standpoint, has proved valid over the decades—as is witnessed by the September 11 suicide attacks by al-Qaeda.

Protracted revolutionary warfare demands a complicated organization on both political and military levels. Mao early developed a clandestine political-military hierarchy that began with the cadre or cellular party structure at the hamlet-village level and proceeded to the top via district, province, and regional command structures. This was roughly the concept followed by guerrilla forces in Malaya and Indochina. Tito was careful to build a parallel political organization in areas that came under his control as a foundation for his future government. Other guerrilla leaders formed civil organizations to provide money, supplies, intelligence, and propaganda. The Viet Cong, Algerian rebel groups, and the Palestine Liberation Organization (PLO) established provisional governments in order to win international recognition, financial backing, and in some instances recognition by the United Nations.

Divisions within political and military commands stemming from ego, envy, ambition, greed, and ignorance have plagued guerrilla leaders through the centuries and are probably more responsible for failed insurgencies than any other factor. The Algerian rebellion of the 1950s suffered severely until the National Liberation Front either absorbed or neutralized rival guerrilla groups, but it failed to settle feuds between the Arabs and the Berbers or between its own internal and external commands. Colombian rebel groups are frequently in conflict. The IRA lost a great deal of effectiveness when it splintered in 1969. Chechnyan rebels are divided between Islamic extremists, who insist on gaining an independent state ruled by Islamic Sharīʿah law, and Russian Orthodox guerrilla fighters, including those who favour an autonomous government under Russian rule. The Tamil Tigers in Sri Lanka are believed to have been divided between Prabhakaran's hard-liners, who demand a separate state, and moderates, who want peace and would accept a reasonable autonomy. At least three major rebel groups and numerous splinter groups are at work in the Philippines, including Islamic fundamentalists, moderate Muslims, and communists. During the Afghan War against Soviet occupation in the 1980s, a score or more of mujahideen rebel groups, ranging from a few hundred to several thousand fighters, were held precariously together by the Islamic religion, an infusion of several billion U.S. dollars, enormous profits from the opium trade, and the desire of each warlord to enlarge his traditional turf. Scarcely had the Taliban government been overthrown by U.S. and allied forces in late 2001 than the warlords turned on one another and on the newly established central government, creating a dangerous semi-anarchy.

Arms

The guerrilla by necessity must fight with a wide variety of weapons, some homemade, some captured, and some supplied from outside sources. In the early stages of an insurgency, weapons have historically been primitive. The Mau Mau in Kenya initially relied on knives and clubs (soon replaced by stolen British arms). French and American soldiers in Vietnam frequently encountered homemade rifles, hand grenades, bombs, booby traps, mines, and trails studded with *punji* stakes soaked in urine (to ensure infection). Nearly every guerrilla campaign has relied on improvisation, both from necessity and to avoid a cumbersome logistic tail. Molotov cocktails and plastique (plastic explosive) bombs are cheap, yet under certain conditions they are extremely effective. Stolen and captured arms also traditionally have been a favourite source of supply, not least because army and police depots also stock ammunition to fit the weapons.

The worldwide proliferation of weapons during the decades of the Cold War added a new dimension to guerrilla capabilities, as the superpowers and other states provided modern assault rifles, machine guns, mortars, and such sophisticated weapons as rocket-propelled grenades and antitank and antiaircraft missiles. The collapse of the Soviet Union and the transformation of some of its republics into independent states brought on a fire sale of more weapons. Many other weapons, however, also came from the busy arsenals of the United States, the United Kingdom, France, Germany, Russia, and Israel.

This largesse has proved to be a double-edged sword for rebels. Although it has improved their staying power, it has also produced an unwanted financial and logistic requirement to feed the hungry weapons and at times has led to quasi-conventional set-piece battles—usually to the guerrillas' regret.

Sanctuary and Support

It was axiomatic to Mao and his followers that revolution begins in familiar terrain. Once sufficient base and operational areas are established, guerrilla operations can be extended to include cities and vulnerable lines of communication. This rural strategy may be influenced by such factors as political goal, geography, and insurgent and government strengths.

If a guerrilla force is to survive, let alone prosper, it must control safe areas to which it can retire for recuperation and repair of arms and equipment and where recruits can be indoctrinated, trained, and equipped. Such areas are traditionally located in remote, rugged terrain, usually mountains, forests, and jungles.

Sympathetic neighbouring countries may also provide sanctuary, both as a physical redoubt and as a source of material support. Ho's guerrillas, in the later stages of Vietnam's war against France, relied on China for refuge, training, and supply of arms and equipment; later, in the war against the United States, they used Laos and Cambodia for sanctuary. Still later Thai guerrillas found sanctuary and support in Cambodia, as did Nicaraguan guerrillas in Honduras. Palestinian irregulars have often enjoyed refuge in Arab states bordering Israel, and a wide variety of militant Islamist groups found refuge in Afghanistan during the 1990s. For years the Basque ETA terrorists (of Spain) took cover in France. Islamic terrorists in the Philippines routinely lose themselves in the jungles of small southern islands. Chechnyan guerrillas frequently find sanctuary in the neighbouring Russian republic of Ingushetiya and in the independent state of Georgia.

People offer a final form of sanctuary, one especially important to an urban guerrilla employing terrorist tactics. A sympathetic population can turn a blind eye to guerrilla activity, or it can actively support operations. During the Cypriot war Grivas was surrounded by a British force for nearly two months without being captured. An Algerian rebel leader installed himself within 200 yards of the army commandant's office in Algiers. The position of neither rebel leader was betrayed despite generous inducement offered to collaborators. An outstanding example from more recent times is the disappearance of Osama bin Laden and Taliban leader Mohammed Omar despite an intensive manhunt and a reward of $25 million for information leading to their capture. U.S. forces ultimately killed bin Laden on May 2, 2011, in a raid on his compound in Abbottabad, Pakistan. Not even the Pakistani government was informed of the mission.

Terror

Terror is one of the most hideous characteristics of guerrilla warfare yet one of its most basic and widely used weapons. It is employed on several levels for several reasons. Tactically, its purpose is to intimidate the military-police opposition—for example, by slitting the throat of a careless sentry or by tossing a grenade into a provincial police outpost. At a slightly higher level it is used to eliminate political and military leaders and officials in order to destabilize the government; to persuade the general populace to offer sanctuary, money, and recruits; and to maintain discipline and prevent defections within the organization. On a still higher level it is used to focus attention on the rebel cause with the hope of winning international support (including financing and recruits) while maintaining internal morale.

It is important to note that up to a certain point the use of terror, though condemned by orthodox governments, is expected and is also a major tactic in counterguerrilla warfare. But what is that certain point? Public opinion seems to put it at promiscuous murder, as exemplified by bomb attacks, whether suicidal or otherwise, against civilian targets. In defense of such attacks, terrorists point to their debilitating effect both in destabilizing governments and in bringing on excessive military reprisals that cost the government public support. What guerrillas risk in such attacks, however, is crossing a line that the public draws between guerrilla fighters and common criminals.

Not all guerrilla leaders have favoured the use of such extreme tactics, either because of humanitarian concerns or because they realize that the resultant stigma outweighs the psychological gains. In Palestine the more moderate Haganah broke with two other Zionist militias, Irgun Zvai Leumi and the Stern Gang, over the issue. In Ireland IRA leaders had sharp disagreements on the use of extreme terror, which resulted in a movement divided between "official" and "provisional" wings, along with numerous splinter groups. Although the PLO denounced the use of such tactics, Ḥamās, Islamic Jihad, and al-Aqṣā Martyrs Brigade continued to employ them on the grounds of justifiable retaliation for military terrorism—as did other groups in Chechnya, Spain, the Philippines, and elsewhere, while also using them for purposes of intimidation and identification.

It is difficult to assess the psychological impact of criminal terrorism on the general population, but it appears that even those persons passively sympathetic to a guerrilla cause are slowly alienated by terrorists planting bombs in shopping centres and holiday resorts or blowing passenger aircraft out of the sky. The sea change in public opinion may have come with the September 11 aerial suicide attacks against American targets and with the United States' subsequent "war on terror." After Sept. 11, 2001, guerrilla warfare, no matter the form or purpose, was generally judged by Western and some Eastern countries to be anathema. Law-enforcement agencies and military forces around the globe were enlarged and adapted to fight terror—literally and with no holds barred. The unforeseen results have been several, but the most unfortunate one has been the use of the war on terror as a shield for continuing abuses by the military, paramilitary, or police in fighting domestic insurgencies. The result is ironic: the more repressive the military terrorism, the greater the number of moderates who come to sympathize with extremists and turn a blind eye to their murderous attacks—a vicious cycle sadly illustrated in the Israeli-Palestinian conflict as well as conflicts in Sri Lanka, the Philippines, Chechnya, Indonesia, Northern Ireland, and elsewhere.

STRATEGY AND TACTICS

The broad strategy underlying successful guerrilla warfare is that of protracted harassment accomplished by extremely subtle, flexible tactics designed to wear down the enemy. The time gained is necessary either to develop sufficient military strength to defeat the enemy forces in orthodox battle (as did Mao in China) or to subject the enemy to internal and external military and political pressures sufficient to cause him to seek peace favourable to the guerrillas (as the Algerian guerrillas did to France, the Angolan and Mozambican guerrillas to Portugal, and the North Vietnamese and Viet Cong to the United States). This strategy embodies political, social, economic, and psychological factors to which the military element is often subordinated—without, however, lessening the ultimate importance of the military role.

That role varies greatly, as does the way it is carried out. T.E. Lawrence's Arabian campaign (1916–18) was strategically vital in protecting the flank of the British general Edmund Allenby's conventional army during its advance in Palestine, yet its success hinged on carrying out the Arabs' political aim, which was to expel Ottoman forces from tribal lands. Lawrence's acceptance of this goal, combined with his linguistic ability, imagination, perception, and immense energy, helped him to establish and maintain unity of command. Popular support was ensured in part by tribal loyalties and hatred of the Ottomans, in part by effective propaganda and decent treatment of the people. There were too many Ottoman soldiers to risk doing battle, but in any case killing the enemy was secondary to killing his line of communication. In Lawrence's words (published in his classic account *The Seven Pillars of Wisdom* [1935]), "the death of a Turkish bridge or rail" was more important than attacking a well-defended garrison. Lawrence kept discipline and organization (Arab, not Western, style) simple and effective. He drilled his men in the employment of light machine guns and in rudimentary demolitions. Camels provided transport. The terrain was desert and desert was sanctuary, and the guerrillas were "an influence, a thing invulnerable, intangible, without front or back, drifting about like a gas." Demanding "perfect intelligence, so that plans could be made in complete certainty," Lawrence "used the smallest force in the quickest time at the farthest place." Mobility and surprise were everything. Hit-and-run tactics on a broad front cut communication, eventually causing enemy garrisons to wither on the vine. By war's end the Arabs had gained control of some 100,000 square miles (259,000 sq. km) while holding 600,000 Ottoman soldiers in passive defense. Arabs had killed or wounded 35,000 enemy at little loss to themselves. They had protected Allenby's vital flank in Palestine and had proved the truth of Lawrence's later dictum: "Guerrilla warfare is more scientific than a bayonet charge."

Mao's political goal was the communist takeover of China. Guerrilla warfare alone, he realized, could not achieve this, but in a prolonged war it was an indispensable weapon, particularly in holding off the enemy (Chinese and Japanese) until orthodox armies could take to the field.

Mao's guerrilla campaign of over two decades stressed the flexible tactics based on surprise and deception that the ancient writer Sunzi had called for in *The Art of War*. Mao later wrote that "guerrilla strategy must be based primarily on alertness, mobility, and attack." He demanded tactics based on surprise and deception: "Select the tactic of seeming to come from the east and attacking from the west; avoid the solid, attack the hollow; attack, withdraw; deliver a lightning blow, seek a lightning decision." Mao instructed his subordinates to accept battle only under favourable conditions, otherwise avoid it and retreat: "We must observe the principle, 'To gain territory is no cause for joy, and to lose territory is no cause for sorrow.' " Careful planning was vital: "Those who fight without method do not understand the nature of guerrilla action."

Ho and his able military commander Vo Nguyen Giap were disciples of Mao's teachings, as was shown in their remarkably successful campaigns against the French and, later, against the U.S. and South Vietnamese armies. Ho and Giap did not, however, hesitate to extend guerrilla operations to the cities when occasion warranted. Vietnamese organization and leadership were generally effective, albeit expensive in lives. The use of terrain was often masterful, both tactically and for sanctuary. When popular support lagged, terrorist tactics were used—particularly the murder of pro-government village headmen—to coerce peasants into furnishing recruits, food, and information while denying these to the enemy. Operations were carefully planned and audaciously executed. As cruel as it was, the guerrilla portion of the Indochina wars must rank as one of the most successful in history.

Leaders who do not respect the principles of guerrilla warfare soon find themselves in trouble, particularly against effective counterguerrilla forces. Greek communist guerrillas lost their war (1946–49) for a variety of reasons, not so much because Tito deprived them of sanctuary in and supply from Yugoslavia but more because they forfeited popular support in northern Greece by their barbarous treatment of civilian hostages, by their rapacious behaviour in villages, and by kidnapping children and sending them to be raised in communist countries.

Filipino, Malayan, and Indonesian guerrillas of the 1940s and '50s suffered from poor organization and leadership as well as from lack of external support, and later movements failed for similar reasons. Uruguayan and Guatemalan insurgents lost control over terrorist tactics and suffered heavily for it. Basque guerrillas, who wanted independence from Spain because of their distinctive language and culture, became unpopular in Spain because of their brutal assassinations. Polisario fighters, inadequately supported by Algeria and Libya, faced continuing stalemate in their war against Morocco over Western Sahara. Angolan and Mozambican guerrillas split into several factions and became pawns of Cuba (and by extension the Soviet Union), South Africa, and the United States. The use of indiscriminate terrorist tactics by the provisional wing of the IRA brought general opprobrium on their movement, including a partial loss of what had been

Lawrence of Arabia

When World War I began in August 1914, T.E. Lawrence became a civilian employee of the Map Department of the War Office in London, charged with preparing a militarily useful map of Sinai. By December 1914 he was a lieutenant in Cairo. Experts on Arab affairs—especially those who had traveled in the Turkish-held Arab lands—were rare, and he was assigned to intelligence, where he spent more than a year, mostly interviewing prisoners, drawing maps, receiving and processing data from agents behind enemy lines, and producing a handbook on the Turkish army. When, in mid-1915, his brothers Will and Frank were killed in action in France, Lawrence was cruelly reminded of the more active front in the West. Egypt at the time was the staging area for Middle Eastern military operations of prodigious inefficiency; a trip to Arabia convinced Lawrence of an alternative method of undermining Germany's Turkish ally. In October 1916 he had accompanied the diplomat Sir Ronald Storrs on a mission to Arabia, where Ḥusayn ibn 'Alī, emir of Mecca, had the previous June proclaimed a revolt against the Turks. Storrs and Lawrence consulted with Ḥusayn's son Abdullah, and Lawrence received permission to go on to consult further with another son, Fayṣal, then commanding an Arab force southwest of Medina. Back in Cairo in November, Lawrence urged his superiors to abet the efforts at rebellion with arms and gold and to make use of the dissident sheikhs by meshing their aspirations for independence with general military strategy. He rejoined Fayṣal's army as political and liaison officer.

Lawrence was not the only officer to become involved in the incipient Arab rising, but from his own small corner of the Arabian Peninsula he quickly became—especially from his own accounts—its brains, its organizing force, its liaison with Cairo, and its military technician. His small but irritating second front behind the Turkish lines was a hit-and-run guerrilla operation, focusing upon the mining of bridges and supply trains and the appearance of Arab units first in one place and then another, tying down enemy forces that otherwise would have been deployed elsewhere, and keeping the Damascus-to-Medina railway largely inoperable, with potential Turkish reinforcements thus helpless to crush the uprising.

Aqaba—at the northernmost tip of the Red Sea—was the first major victory for the Arab guerrilla forces; they seized it after a two-month march on July 6, 1917. Thenceforth, Lawrence attempted to coordinate Arab movements with the campaign of General Sir Edmund Allenby, who was advancing toward Jerusalem, a tactic only partly successful. In November Lawrence was captured at Darʿā by the Turks while reconnoitring the area in Arab dress; he was apparently recognized and homosexually brutalized before he was able to escape. The experience, variously reported or disguised by him afterward, left real scars as well as wounds upon his psyche from which he never recovered. The next month, nevertheless, he took part in the victory parade in Jerusalem and then returned to increasingly successful actions in which Fayṣal's forces nibbled their way north. Lawrence rose to the rank of lieutenant colonel with the Distinguished Service Order (DSO).

By the time the motley Arab army reached Damascus in October 1918, Lawrence was physically and emotionally exhausted, having forced body and spirit to the breaking point too often. He had been wounded numerous times, captured, and tortured; had endured extremities of hunger, weather, and disease; had been driven by military necessity to commit atrocities upon the enemy; and had witnessed in the chaos of Damascus the defeat of his aspirations for the Arabs in the very moment of their triumph, their seemingly incurable factionalism rendering them incapable of becoming a nation. (Anglo-French duplicity, made official in the Sykes-Picot Agreement, Lawrence knew, had already betrayed them in a cynical wartime division of expected spoils.) Distinguished and disillusioned, Lawrence left for home just before the Armistice and politely refused, at a royal audience on Oct. 30, 1918, the Order of the Bath and the DSO, leaving shocked King George V (in his words) "holding the box in my hand." He was demobilized as a lieutenant colonel on July 31, 1919.

heavy financial support from previously sympathetic Irish Americans.

Why then do guerrilla leaders condone criminal terrorism? Not all of them are able to prevent its use, but, as is mentioned above, terrorist campaigns have played and continue to play important roles in forcing reluctant governments into negotiations. Negotiation, however, is not to the taste of some guerrilla leaders, especially those who reckon that their demands are being unfairly pared down. The discontented are usually extremists who may take their followers and splinter from the main group in order to continue their own war. In some cases they will be financed by outside agencies, such as extremist religious organizations, or by selling their services to criminal organizations, as has happened in Colombia, Northern Ireland, and Spain. Splinter groups may also find support at home, depending on the kind of campaign conducted against them by the government.

COUNTERGUERRILLA WARFARE

Perhaps the most important challenge confronting the military commander in fighting guerrillas is the need to modify orthodox battlefield thinking. This was as true in ancient, medieval, and colonial times as it is today. Alexander the Great's successful campaigns resulted not only from mobile and flexible tactics but also from a shrewd political expedient of winning the loyalty of various tribes (Alexander recruited one guerrilla leader into his army and then married his daughter). The few Roman commanders in Spain—Tiberius Sempronius Gracchus, Marcus Porcius Cato, Scipio Africanus the Elder and the Younger, and Pompey the Great—who introduced more mobile and flexible tactics often succeeded in defeating large guerrilla forces, and their victories were then exploited by decent treatment of the

U.S. soldiers looking for insurgents in Iraq, March 5, 2005. United States Department of Defense/Airman 1st Class Kurt Gibbons III, U. S. Air Force.

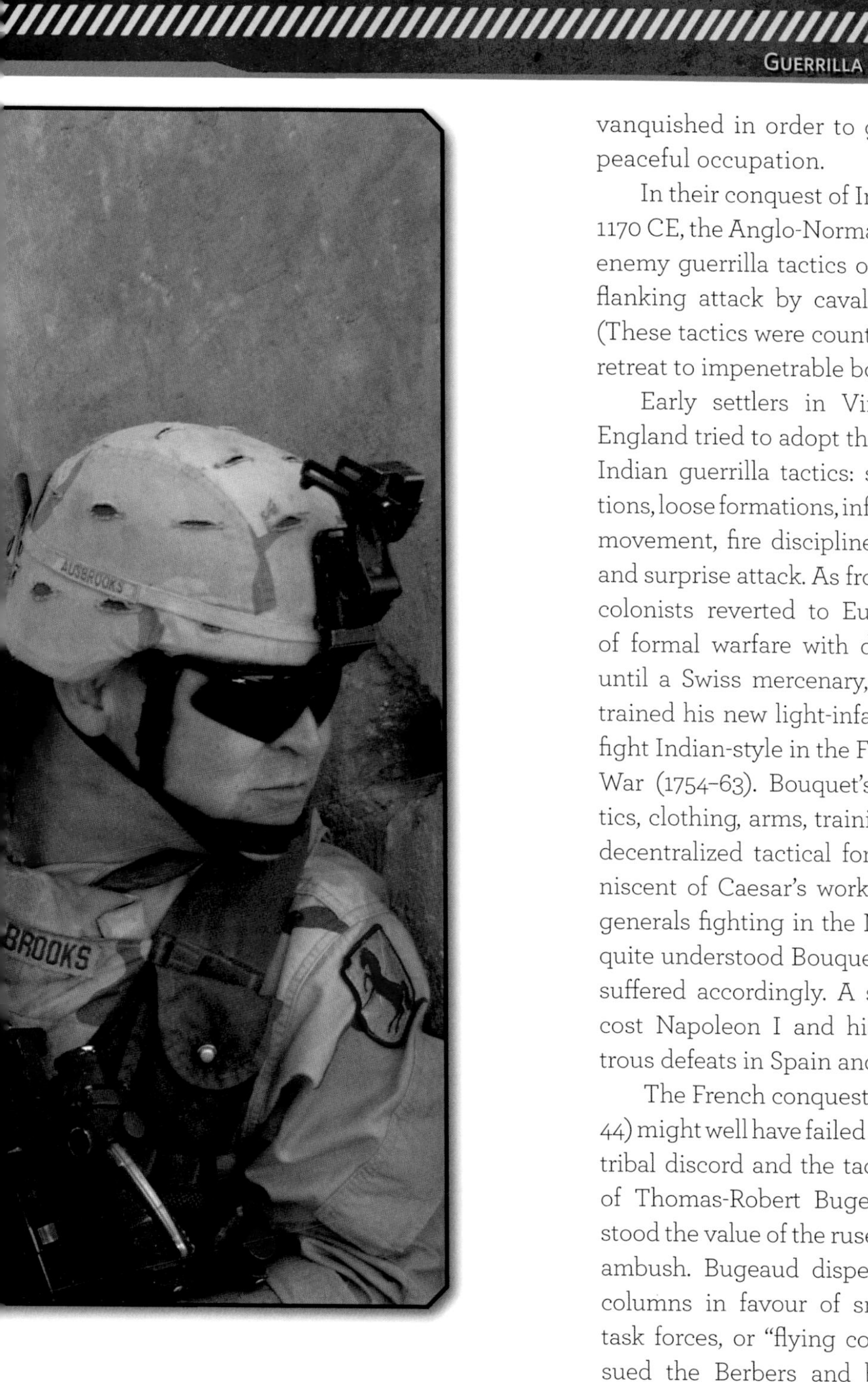

vanquished in order to gain a relatively peaceful occupation.

In their conquest of Ireland starting in 1170 CE, the Anglo-Normans borrowed the enemy guerrilla tactics of feigned retreat, flanking attack by cavalry, and surprise. (These tactics were countered by the Irish retreat to impenetrable bog country.)

Early settlers in Virginia and New England tried to adopt the best features of Indian guerrilla tactics: small-unit operations, loose formations, informal dress, swift movement, fire discipline, terror, ambush, and surprise attack. As frontiers expanded, colonists reverted to European methods of formal warfare with disastrous results until a Swiss mercenary, Henry Bouquet, trained his new light-infantry regiment to fight Indian-style in the French and Indian War (1754–63). Bouquet's treatise on tactics, clothing, arms, training, logistics, and decentralized tactical formations is reminiscent of Caesar's work on Gaul. British generals fighting in the New World never quite understood Bouquet's teachings and suffered accordingly. A similar blindness cost Napoleon I and his generals disastrous defeats in Spain and Russia.

The French conquest of Algeria (1830–44) might well have failed had it not been for tribal discord and the tactical innovations of Thomas-Robert Bugeaud, who understood the value of the ruse, the raid, and the ambush. Bugeaud dispensed with heavy columns in favour of small, fast-moving task forces, or "flying columns," that pursued the Berbers and brought them to battles that were usually won by disciplined French troops using superior arms.

Although Bugeaud believed in constructive occupation—"the sword only prepared the way for the plough"—he nonetheless depended more on fear than on persuasion, relying on the *razzia* (raid) to implement a scorched-earth policy to starve the natives into submission. Bugeaud's offensive tactics of clearing, holding, and expanding became the model for subsequent pacification campaigns around the globe, including the United States' winning of the West and U.S. forays into colonialism in Cuba and the Philippines.

Such were the string of colonial successes that occasional serious reverses due to inept leadership and ill-trained troops were shrugged off. Orthodox commanders were generally quite content to put unquestioning faith in sheer military weight with little consideration given either to the poor organization and leadership of native forces or to the lack of modern arms and allies. Blockhouses and garrisons kept the peace in pacified areas. If natives rebelled, they were put down with force.

This simplistic approach was challenged by a French general, Louis-Hubert-Gonsalve Lyautey. He had been taught by Joseph-Simon Gallieni in Indochina in 1895 (the French had gradually been asserting control over that region since the late 1850s) that military success, in Gallieni's words, meant "*nothing* unless combined with a simultaneous work of organization—roads, telegraphs, markets, crops—so that with the pacification there flowed forward, like a pool of oil, a great belt of civilization." Lyautey later employed this *tache d'huile*, or oil-spot, strategy in Algeria, where he used the army not as an instrument of repression but, in conjunction with civil services, as a positive social force—"the organization on the march."

Lyautey's success went generally unheeded—the South African War (1899–1902), for instance, introduced the use of the concentration camp for Boer civilian noncombatants. Native rebellions continued to be put down with force; orthodox commanders were not greatly impressed with the guerrillas in World War II. The greater was the postwar shock, then, when these commanders and their subordinates were called upon to quell organized insurgencies by ideologically motivated, combat-trained guerrillas equipped with modern weapons and often politically allied with and supplied by the Soviet Union and its satellite countries.

Most governments and commanders simply floundered while calling for more soldiers and more weapons. The Greek army originally tried to suppress what they termed "bandits" by using static defense tactics that soon failed. Once the army had received massive reinforcements of U.S. arms and equipment, it launched large-scale offensives, or "search-and-clear" operations, which met with only limited success. Chinese Nationalist commanders moved vast armies hither and yon in futile efforts to capture Mao's guerrillas before finally holing up in towns and cities, where they eventually fell prey to Mao's own army divisions. During the Hukbalahap Rebellion (1946–54), U.S. Army advisers in the Philippines trained and equipped Filipino combat teams supported by armour, aircraft, artillery, and even trained war dogs. Large-scale search-and-destroy operations—the "ring of steel" tactic similar to that unsuccessfully employed by German commanders against Tito's Yugoslavian guerrillas—produced minimal results, as did free-fire areas (zones in which troops may fire at anything and everything), massive and sometimes brutal interrogations of villagers, and the employment of terrorist tactics, all of which further alienated the rural people whose support was necessary to defeat the guerrillas. Wiser commanders replaced conventional tactics with small-unit patrols and a variety of ruses that largely neutralized overt guerrilla action, then turned the army to the vital task of winning civil cooperation. With this the Huk insurgency died, but by the 1970s the failure to carry out promised reforms, mainly land distribution, brought on a guerrilla insurgency by the New People's Army that lasted into the 21st century.

British commanders in Malaya also performed ineffectually in the early phases of the communist insurgency that began in 1948. Eventually, however, they realized that the support of the rural natives was vital to their goal of eliminating the entire guerrilla apparatus. Once they had achieved a reasonable civil-military chain of command, their first priority became the reestablishment of law and order, which meant revitalizing the rural police function. The military effort concentrated on breaking up and dispersing large guerrilla formations, then depriving them of the initiative by small-unit tactics—mainly frequent patrols and ambushes based on valid intelligence often gained from natives. The subsequent civil effort was designed to win "the hearts and minds" of the people, first by providing security in the form of village police and local militias working with government forces, second by providing social reforms (land reform, schools, hospitals) that identified the government with the people's best interests. Harsh measures were necessary: the compulsory census, an identity-card system, suspension of habeas corpus (with carefully publicized safeguards), searches of private property without a warrant, the death sentence for persons caught with unauthorized weapons, harsh sentences for collaborators, curfews, resettlement of entire villages, and other extraordinary measures. These were somewhat palliated by the British government's promise of eventual independence and by the general unpopularity of the guerrillas among the majority Malay population as well as among the urban Chinese business community.

American military forces began to recognize the rising importance of unconventional warfare during the Cold War, though this recognition came only grudgingly to the top command. In the early 1950s U.S. Army Special Forces units—later known as the "Green Berets"—were formed as deep-penetration teams designed to contact and support indigenous guerrilla groups in rising against communist governments. Though superbly trained, they suffered from severe linguistic limitations and in the event were never committed. In a notable role reversal during the Vietnam War, numerous Green Beret teams were assigned to assist Montagnard tribes (hill tribes) in countering the generally effective operations of Viet Cong guerrillas—though not with outstanding success in spite of heavy financial and material support.

Orthodox senior commanders in Vietnam and later conflicts seemed oblivious to lessons learned in Malaya and the Philippines, the foremost of which was to offer the opponents, and particularly their supporters, a government that would fairly adjudicate their grievances. Believing solely in a military victory, they relied on tactics that only further alienated the very people whose hearts and minds had to be won over if the guerrillas were to be denied their support. Wholesale aerial bombings, mass artillery interdiction of suspected sanctuary areas, division- and corps-strength "sweeps" in which few guerrillas were captured or killed while entire villages were destroyed, free-fire areas that resulted in the deaths of women and children, isolated chains of military outposts and static defensive barriers that were easily outflanked, mass arrests, brutal interrogations, and cruel incarcerations—all of these amounted to a frightful expenditure of lives and money as one country after another threw in the towel, the United States in Vietnam, France in Algeria, and the Soviet Union in Afghanistan.

These campaigns failed on two levels. On the civil level, the authorities refused to admit the validity of often well-founded grievances and failed to undertake vital and generally long-overdue reforms under military and police protection for as long as was necessary. On the military level, the specific failures cited above can be summarized in four words: too much too soon. In order to be successful, counterguerrilla warfare must be a happy marriage between civil and military authority, between the civilian administrator and the soldier-policeman. For the administrator to function properly, the rebels must be contained and then neutralized—a long and arduous task. Throughout history commanders have proudly pronounced the demise of the guerrilla only to witness his reappearance in a year or two, as happened in Peru with the Sendero Luminoso ("Shining Path") group.

The key to waging successful counterinsurgency warfare lies in the nature of the insurgency. If an insurgency is an ill-founded uprising, either political or criminal, a legitimate government can treat it as such and can call on the support of other governments if necessary. But if an insurgency is founded on legitimate grievances that an ineffectual, biased, or corrupt government refuses to recognize, much less amend, then the conflict will not be ended until that government agrees to reach a solution by negotiation, not force. Too many governments, influenced by strong military establishments or by sweeping declarations of war, have refused to recognize the legitimacy of guerrilla challenges, seeking instead an ephemeral victory by means of military force, which is eventually answered in kind by guerrilla warfare.

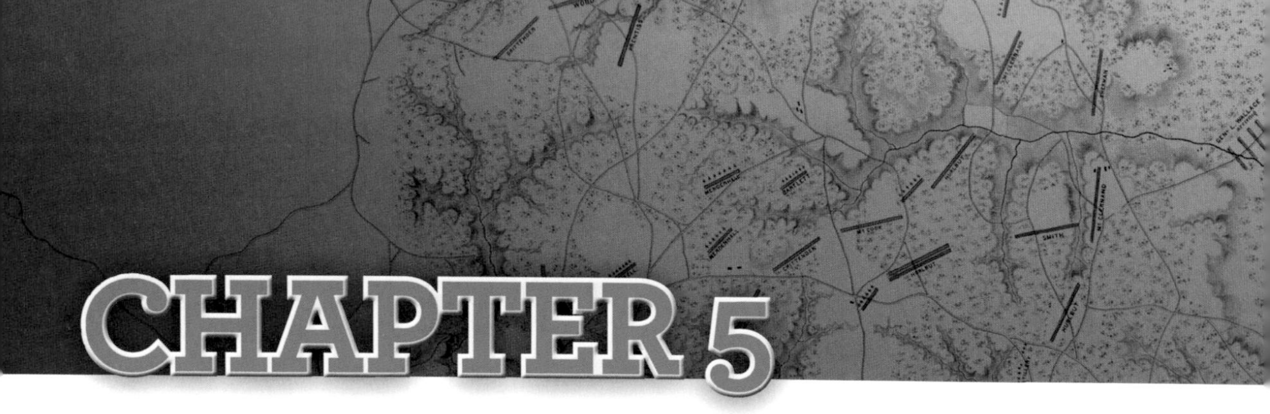

CHAPTER 5

TOPICS IN 21ST-CENTURY WARFARE

As the 20th century closed, it seemed to some historians, of military science as well as politics and society, that history as traditionally understood had ended. Gone were the mass movements and trends of past centuries—identifiable quantities that could be reduced to first principles in a true science of war. Taking their place were new uncertainties. One uncertainty was the diverse aspirations of developing nations around the world. Another was the appearance of constantly changing electronic technologies that were just as revolutionary in their way as the rise of industry in the 19th century. These developments as well as a host of others have made it almost impossible to formulate a single 21st-century science of war. But wars have continued to be fought, and the first decade of the 21st century saw a diverse enough array of changes, from high-tech warfare to child soldiers, to make it clear that the 21st century would be a century of change across the gamut of military sciences, from the classic triad of strategy, tactics, and logistics to the always-shifting and amorphous guerrilla warfare.

THE LIMITS OF HIGH TECHNOLOGY: AFGHANISTAN, 2001–02

The war that began in Afghanistan on Oct. 7, 2001, demonstrated both the capabilities and the limitations of modern military technology. It should have come as no surprise that the U.S.-led 17-member coalition toppled the Taliban regime in only a few weeks. In conventional terms, the Taliban were a pushover; they possessed no air force, had very limited air defenses, and were an unpopular and weak regime. It must be remembered, however, that in 1979 the Soviet Union controlled Afghanistan's capital, Kabul, within a week of beginning its invasion and then spent the next decade trying to defeat the mujahideen guerrillas. The U.S.-led coalition faced a similar challenge against widely dispersed and tenacious al-Qaeda forces operating in rugged and inhospitable terrain. Consequently, the coalition was able to chase al-Qaeda and its Taliban hosts from the field in 2001, but it was not able to eliminate al-Qaeda's terrorist infrastructure.

New Weapons

The limitations of high technology were quickly demonstrated in the use of new weapons. Shortly after the war began, an American bomb designed to destroy underground tunnels and bunkers was rushed into service. The BLU-118/B thermobaric bomb was dropped on a suspected enemy cave in the eastern part of the country in March 2002. Unfortunately, although the device detonated as intended, creating a large fireball and a devastating shockwave, a problem with its laser guidance caused it to land far enough away from the cave entrance to negate its effect. (Ironically, the Soviets also employed thermobaric bombs in Afghanistan in the 1980s.)

Laser guidance was a revolution in air warfare that was adopted by all advance military forces, but it had two main disadvantages: the laser beam marking the target had to be aimed by someone on the ground or in an aircraft, and smoke and bad weather could degrade the laser beam such that it could no longer guide the falling bomb. In Afghanistan such problems were quickly addressed through various technologies, including the use of the Global Positioning System (GPS). A computer mounted in the bomb was programmed with the coordinates of the intended target and used GPS guidance to strike its target. Since the 1990–91 Persian Gulf War, special forces troops had been training to use handheld GPS receivers, laser designators, and satellite radios to help artillery and aircraft attack targets with minimal delay. This capability assumed even greater importance in Afghanistan, where reducing the "sensor to shooter" loop to just a few minutes was necessary to pin down and destroy small groups of guerrillas on the move.

The Afghanistan War would forever be remembered as the first in which armed unmanned aerial vehicles (UAVs) were used to attack targets. UAVs had been in service for more than 40 years as drones for target practice and to gather intelligence with onboard sensors, but early in the Afghanistan War they came into use as weapons platforms, and as such they immediately began to compile a record of successes—punctuated by failures. For instance, in February 2002 the CIA used a Predator UAV to fire an antitank missile at a group of three men believed to be al-Qaeda leaders. All were killed, but it later turned out that they were local villagers. The costs and benefits of using UAVs was to become an important issue in the war.

LOGISTIC CHALLENGES

What set the U.S. military apart from all others at the turn of the 21st century was its ability to dispatch thousands of troops and their weapons, vehicles, and supplies to any point on Earth and to sustain them there. No other country could wage war in a landlocked country by supplying its forces almost entirely by air. By the end of September 2001, nearly the entire active-duty U.S. fleet of C-5 Galaxy and C-17 Globemaster III transport aircraft—a total of about 140—was dedicated to the war effort. The 30-year-old C-5 could carry 122,000 kg (270,000 pounds) of cargo, but it required a runway at least 1,500 metres (4,900 feet) long for landing. Conversely, the C-17 could land on runways as short as 915 metres (3,000 ft), which made it much better suited to the primitive or war-damaged airfields in Afghanistan. In early 2002, coalition troops each month consumed 7.9 million litres (2.1 million gallons) of fuel, 13.6 million litres (3.6 million gallons) of water, and the equivalent of 72 18-wheel transport trucks of food. Meeting such a demand for supplies did not come cheap, however. For example, the price of delivering fuel to remote war zones exceeded $1,500 per litre.

REMOTE COMMAND AND INFORMATION OVERLOAD

For the most part, the invasion of Afghanistan was directed from U.S. Central Command headquarters in Florida, more than 11,260 km (7,000 miles) and 10 time zones away. Commanders for the first time were able to watch battles live via television cameras mounted in UAVs. Although an impressive technical achievement, this led to complaints that the attention of headquarters staff was diverted and that troops in the field were being micromanaged.

The large volume of data moving between commanders and troops in the field also was a mixed blessing for coalition forces. On one hand it allowed commanders to deploy forces quickly and effectively to where they were needed most, but on the other hand information overload created a requirement for new staff positions, such as "knowledge management officer" to filter out minor details and ensure that commanders got only the information they needed in order to make decisions.

WAR AND THE MEDIA: IRAQ, 2003

On March 20, 2003, Anglo-American ground forces crossed into Iraq in order to overthrow Pres. Saddam Hussein. The U.S.-led coalition's campaign against Hussein was an entirely new experience, not only for the fighting troops but also for the reporters and crews covering the action. The Iraq invasion was the first sustained, conventional land campaign for many years to be fought by troops from major Western democracies. In the Persian Gulf War of 1990–91, ground troops were in battle for just four days, while in Kosovo and Afghanistan—at least to that time—ground forces from Western countries were not involved to any significant degree.

Since 1991, however, technology had transformed the way the media worked. By 2003 satellite communications had become compact, mobile, and cheap; 24-hour television and radio news channels had become familiar throughout the world; and the Internet offered the capacity to deliver news around the globe just minutes after it had been written. Live reports could be transmitted from almost every battle zone, so the public could follow the invasion, or at least some aspects of it, virtually in real time. This resulted in some powerful images and pieces of reporting, both from the front lines of the coalition forces and from inside Iraqi cities, and, for the first time, from Arab as well as Western sources. The Qatar-based Al-Jazeera television station had access to Basra and parts of Baghdad from which Western journalists were barred until those two cities were occupied by coalition forces.

"Embedded" Versus "Unilateral" Journalists

Journalists who wanted to report on the fighting from the front lines had two options. They could become "embedded" with coalition military units or operate independently as "unilaterals." Some 600 journalists, about 450 of them from the United States, chose to be "embeds." Each lived with his or her unit and held the honorary rank of major. They witnessed the invasion firsthand, with almost complete access to the troops. In return, they agreed not to write about imminent attacks, future operations, or classified weapons. Journalists also agreed to report on military actions in only general terms to prevent Iraqi forces from securing vital intelligence. (Geraldo Rivera of Fox News was temporarily removed from his unit for revealing its exact position.)

The embedded journalists produced many dramatic firsthand reports of the fighting as the coalition forces advanced on Baghdad, but doubts surfaced about their ability to assess the wider progress of the invasion. On March 26 several "embeds" reported that a convoy of up to 120 Iraqi tanks was leaving Basra. The next day a British spokesman admitted that only 14 tanks had left the city. In addition, the very status of these embedded journalists might have compromised their independence. Phillip Knightly, the Australian-born author of *The First Casualty*, one of the standard books on the history of war reporting, said, "I was able to find only one instance of an embedded correspondent who wrote a story highly critical of the behaviour of U.S. troops." This was when William Branigin of the *Washington Post* reported the deaths of Iraqi civilians at a U.S. military checkpoint. The official account said that

Naval aviators briefing embedded media members during a press conference aboard the aircraft carrier USS Harry S. Truman, *March 22, 2003, in the Mediterranean Sea en route to the war in Iraq.* U.S. Navy photo by Photographer's Mate 1st Class Michael W. Pendergrass

warning shots had first been fired at a car that refused to stop. Branigin wrote that no such shots were fired.

The "unilaterals" had fewer constraints than their embedded colleagues but also far less access to coalition troops; thus, their ability to report on the fighting proved to be no greater. One of the most significant false stories of the invasion—that, after 10 days, the U.S. forces were planning a pause in their advance on Baghdad—emanated from a group of unilaterals.

Managing Media Access

The coalition established an official press centre at its central command in Qatar, where regular briefings were given to the world's media. Many journalists, however, complained that little useful information was provided. The head of communication planning at Britain's Ministry of Defence subsequently admitted severe shortcomings, including the failure to provide adequate "context-setting briefings."

Although the information provided in Qatar was generally accurate, if sparse, there were times when the fog of war obscured the truth. On April 2 reporters were shown military video film of the rescue of U.S. Army Pvt. Jessica Lynch from an Iraqi military hospital near Nassiriya. According to the official account, which was widely reported around the world, Lynch was part of a maintenance team that had been ambushed on March 23. Nine of the team were killed; Lynch was stabbed and shot but continued to fire back at the Iraqi troops. After she was captured, she was harshly interrogated and slapped about the head. Eight days later U.S. special forces fought their way into the hospital against heavy resistance and rescued her.

Key parts of this account were later found to be untrue. Lynch was wounded but not shot or stabbed, and another soldier in the unit (not Lynch) had fired back. Far from being badly treated in the hospital, she received the best treatment that the Iraqi doctors and their meagre resources could provide. By the time the U.S. special forces arrived, Iraqi troops had left the area. There was no resistance. Moreover, the Iraqi doctors had tried to hand Lynch back to the U.S. Army two days earlier, but when the Iraqi ambulance approached the American lines, U.S. troops opened fire and forced it to turn around.

If the quality of information available to journalists on and behind the coalition lines was variable, it was no better on the other side. Hussein's regime provided no media access to Iraqi troops south of the capital, but it sought to have its side of the arguments—political, diplomatic, and military—conveyed to the outside world via journalists who remained in Baghdad and Basra, notably those working for Al-Jazeera. Although the American television networks withdrew from Baghdad shortly before the start of the war, the BBC and other British broadcasters remained, as did television teams from many other countries. Some major American newspapers, such as the *New York Times,* had correspondents in Baghdad throughout the war.

Iraqi officials—most notably the perennially optimistic information minister, Muhammad Sa'id al-Sahaf, dubbed "Comical Ali" by bemused foreign reporters—consistently denied that the coalition forces were gaining ground. As late as April 9, Sahaf was predicting a comprehensive Iraqi victory, even as U.S. tanks could be seen behind him crossing the Tigris River in the heart of Baghdad.

Iraq imposed no formal censorship on foreign journalists, and live reports were a regular feature from the roof of the Palestine Hotel, the de facto Baghdad headquarters of the international press. Some self-censorship, however, was inevitable. Most Western journalists, especially television crews, employed local staff as fixers, interpreters, and support staff and sought to protect them. Only when central Baghdad fell to coalition forces on April 9 did foreign journalists in the city feel able to abandon such restraint and to report without inhibition.

WINNERS AND LOSERS

The biggest media winners of the invasion of 2003 were the television news channels, which saw their audiences increase dramatically. In the United States, Fox News increased its audience fourfold to a daily average of 3.3 million viewers, overtaking the well-established CNN (with a daily average of 2.65 million viewers). Fox benefited from taking a firmly pro-coalition stance toward the invasion, while CNN upheld its tradition of striving for objective detachment.

Altogether, more than a dozen journalists lost their lives covering the invasion, some of them victims of "friendly fire." The dead included NBC TV's David Bloom, Michael Kelly of the *Washington Post,* Terry Lloyd from Britain's Independent Television News, Christian Liebig of the German magazine *Focus,* Julio Anguita Parrado of the Spanish newspaper *El Mundo,* and Argentine television's Mario Podesta. In Baghdad two cameramen, one with Reuters and one with Spanish television, were killed when U.S. tanks fired at the Palestine Hotel, and an Al-Jazeera correspondent died when at least one U.S. bomb hit the station's Baghdad offices. The International Press Institute criticized the U.S. forces for these attacks on civilian targets.

Despite high-tech developments, war correspondents in the new century clearly faced as many challenges and as much danger as those in previous wars ever had.

POWS AND THE WAR ON TERRORISM: IRAQ AND AFGHANISTAN, 2004

In April 2004 photographs showing abuse of detainees by U.S. soldiers at the notorious Abu Ghraib prison in Baghdad began circulating on the Internet and on televised news programs, setting off a new firestorm of criticism around the world against the U.S. occupation of Iraq. The photos, taken by soldiers at the prison, became key to a dozen investigations, including inquiries by both houses of Congress. During the furor that ensued, more evidence came to light that prisoners held by the United States in various locations had been beaten, sexually assaulted, deprived of sleep and medical attention, frightened by dogs, and subjected to other forms of intimidation, humiliation, and abuse. These acts were part of interrogations, supposedly to get prisoners to reveal useful information about terrorist activities. The controversy renewed critical attention on the behaviour of the participants in the global "war on terrorism" declared by the United States in 2001 and in the Iraq War launched in 2003.

ABU GHRAIB PRISON

In March 2003 the United States led its invasion of Iraq to depose Saddam Hussein, and by May U.S. Pres. George W. Bush declared that all major combat operations had ended. By that time, coalition forces were holding more than 7,000 Iraqi prisoners of war (POWs). Some of these prisoners, as well as combatants captured afterward and other Iraqis arrested for numerous offenses, were detained at the Abu Ghraib prison—a facility notorious for brutality under Hussein's regime. There, in the fall of 2003, prisoners were subjected to various forms of physically and psychologically abusive treatment by their American military captors. They were kept naked for days at a time, photographed in that state, and forced to pose in sexually explicit positions. They were also deprived of sleep and threatened with electric shock or with attacks by military dogs. This treatment violated international humanitarian law, specifically the Geneva Conventions, which prohibited the humiliating or degrading treatment of prisoners of war. According to investigators from the International Committee of the Red Cross (ICRC), some of the abuses could be classified as torture and therefore violated not only the Geneva Conventions but also the International Covenant on Civil and Political Rights, the Convention Against Torture, and the Universal Declaration of Human Rights.

As outrage over the images from Abu Ghraib was expressed worldwide, Bush publicly declared his disgust at the treatment of the prisoners. As an indicator that the abuses were taken seriously by the government, the U.S. military initiated court-martial prosecutions against lower-rank soldiers implicated in the mistreatment. Nine were convicted. However, no criminal charges were ever filed against any of the higher-level officials whom many considered to have authorized or encouraged this type of conduct as an interrogation or disciplinary method.

In response to the Abu Ghraib revelations, Congress eventually passed the Detainee Treatment Act, which banned the "cruel, inhuman, or degrading" treatment of prisoners in U.S. military custody. Although the measure became law with Bush's signature in December 2005, he added a "signing statement" in which he reserved the right to set aside the law's restrictions if he deemed them inconsistent with his constitutional powers as commander in chief.

Guantánamo Bay Detention Camp

The treatment of prisoners at Abu Ghraib renewed attention to other combatants detained elsewhere by the United States. As the invasion of Afghanistan began in October 2001, the Bush administration had declared that captured members of the al-Qaeda terrorist organization were "unlawful combatants." As such, they had no right to protection under international law, and furthermore, such persons could be held indefinitely without formal charges under powers that Congress granted the president to fight terrorism. This classification was a first step toward authorizing trial by military (rather than civilian) courts, where normal due process and constitutional protections would not apply. In the past the United States had condemned the use of military courts to try civilians

A member of the U.S. military standing by while two detainees stand inside the fence line at Camp Delta on May 9, 2006, in Guantánamo Bay, Cuba. Mark Wilson/Getty Images

in countries such as Greece and Turkey, but the U.S. government justified its decision in this case by claiming that normal criminal court proceedings could result in a breach of security or give helpful information to those planning terrorist attacks.

The administration said it would apply the Geneva Conventions to soldiers of Afghanistan's deposed Taliban regime, the Islamic fundamentalist faction that had ruled Afghanistan and had harboured al-Qaeda, though it would not grant them formal status as POWs.

In early 2002 the U.S. detention facility at the Guantánamo Bay Naval Base, located on the coast of Guantánamo Bay in southeastern Cuba, began receiving suspected members of al-Qaeda and fighters for the Taliban. Eventually hundreds of prisoners from several countries were held at the camp without charge and without the legal means to challenge their detentions. The Bush administration maintained that it was neither obliged to grant basic constitutional protections to the prisoners,

POWs and the Geneva Conventions

The Geneva Conventions divide all persons in an armed conflict into two categories: combatants and civilians. Combatants are authorized to fight in accordance with the laws of war on behalf of a party to the conflict. Civilians are not authorized to fight but are protected from deliberate targeting by combatants as long as they do not take up arms. Under the Geneva Conventions, parties to an armed conflict have the right to capture and intern enemy combatants as well as civilians who pose a danger to the security of the state. Enemy combatants are not presumed to be guilty of any crime; rather, they are detained to remove them as a threat on the battlefield. The detaining power has the right to punish enemy soldiers and civilians for crimes committed prior to their capture as well as during captivity, but only after a fair trial in accordance with applicable international law.

The Geneva Conventions stipulate that POWs should be tried in a military court unless the existing laws of the detaining power permit trials of its own military personnel in a civil court for the same offense. POWs have the right to defense by a qualified advocate or counsel of their own choice, to the calling of witnesses, and, if they deem it necessary, to the services of a competent interpreter. For example, former Panamanian leader Gen. Manuel Noriega was given a 30-year prison term by a U.S. federal court for drug trafficking and other crimes even though he was brought to trial as a POW captured during the U.S. invasion of Panama in 1989. According to the ICRC, all detainees taken in war are protected by the Geneva Conventions, and violations of the accords may constitute either war crimes or crimes against humanity.

To be considered POWs under the Geneva Conventions, detainees must fall under one of these categories:

1. *Members of the regular armed forces of a party to the conflict or of militias or volunteer corps forming part of such armed forces*
2. *Members of other militias or other volunteer corps, including those of organized resistance movements, as long as they:*

   (a) *are part of an identifiable command structure*
   (b) *have fixed distinctive insignia recognizable at a distance*
   (c) *carry their arms openly*
   (d) *conduct their operations in accordance with the laws of war*

3. *Members of regular armed forces who profess allegiance to a government or an authority not recognized by the detaining power*
4. *Inhabitants of a nonoccupied territory who have spontaneously taken up arms to resist an invading force, provided that they carry arms openly and respect the laws of war.*

The Geneva Conventions state that should any doubt arise as to whether detainees fit these categories, they "shall enjoy the protection of the present convention" until "their status has been determined by a competent tribunal." Also, precedents can be set that expand or reinforce definitions. During the Korean War the United States and its allies treated Chinese detainees as POWs even though the People's Republic of China was not yet recognized diplomatically. Also, Viet Cong guerrillas captured by the United States during the Vietnam War were given POW status despite the fact that they often wore civilian clothing with no insignia and did not carry their arms openly.

since the base was outside U.S. territory, nor required to observe the Geneva Conventions regarding the treatment of prisoners of war and civilians during wartime, as the conventions did not apply to unlawful enemy combatants.

The camp was repeatedly condemned by international human rights and humanitarian organizations—including Amnesty International, Human Rights Watch, and the ICRC—as well as by the European Union and the Organization of American States, for alleged human rights violations, including the use of various forms of torture during interrogations. In response to such criticism, the Bush administration generally insisted that detainees were well cared for and that none of the "enhanced interrogation techniques" employed on some prisoners amounted to torture. Moreover, according to U.S. officials, the use of such techniques had in many cases—e.g., in the interrogation of Khalid Sheikh Mohammed, the alleged mastermind of the September 11 attacks—yielded valuable intelligence on the leadership, methods, and plans of al-Qaeda and other terrorist organizations.

LEGAL UNCERTAINTY

Whether or not al-Qaeda and other detainees met the Geneva Conventions' criteria for POWs, and to what extent the detainees were entitled to legal rights, was an issue of great contention in 2004 and beyond. In June 2004 the U.S. Supreme Court, in *Rasul* v. *Bush*, ruled that terrorism suspects, including the prisoners held at Guantánamo Bay, had a right to file writs of habeas corpus and to request a review of their cases in U.S. federal courts, because Guantánamo Bay was considered legal territory of the United States. The implication of the ruling was that hundreds of foreign national detainees had a legal right to challenge their imprisonment.

In July 2004 the U.S. military began establishing tribunals to determine the status of suspects accused of being unlawful combatants; however, the suspects were permitted only military representatives and not their own personal civilian lawyers. Later that year, however, a U.S. district court ruled that Bush had overstepped his constitutional bounds and improperly skirted the Geneva Conventions in establishing military commissions to try detainees at Guantánamo as war criminals. The decision was overturned in 2005 by a U.S. court of appeals, and the case went to the Supreme Court. In June 2006 the Supreme Court declared, in *Hamdan* v. *Rumsfeld*, that the military commissions established by Bush to hear charges against individuals alleged to have violated the laws of war were unlawful both because they were inconsistent with the American Uniform Code of Military Justice and because they were not "regularly constituted" courts required by the Geneva Conventions. The court identified as a critical defect in the pre-2006 military commissions a rule permitting a commission to consider secret evidence that was not disclosed to the defendant.

Congress responded to this decision by enacting the Military Commissions Act of 2006, which gave the military commissions the express statutory basis that the Supreme Court had found was lacking. The act guaranteed the right of defendants to be present at commission proceedings, but it also denied the federal courts jurisdiction to hear habeas corpus petitions on behalf of foreign detainees. In 2008, however, the Supreme Court, in *Boumediene* v. *Bush*, overturned the latter provision of the law by ruling that foreign detainees did have the right to challenge their detentions in the federal courts. The ruling declared unconstitutional parts of two laws approved by Congress after the September 11 attacks that were designed to allow indefinite detention of suspects and their eventual trial by military commissions. It further complicated dozens of pending combatant cases that were already burdened with charges of torture, withholding of evidence, and violations of international law by the U.S. military.

Political Uncertainty

Uncertainty in the judicial realm was matched by uncertainty in the political realm. On Jan. 22, 2009, newly inaugurated Pres. Barack Obama fulfilled a campaign pledge by ordering the closure of the facility at Guantánamo within one year and a review of ways to transfer detainees to the United States for imprisonment or trial. He also required interrogators to use only the techniques contained in the U.S. Army's field manual on interrogation, none of which was considered torturous.

The Guantánamo deadline was controversial and in fact proved overambitious. It implied a rejection of the Bush-backed military tribunals conducted outside U.S. soil, suggesting instead that the remaining detainees held at Guantánamo as "enemy combatants" would be either released, transferred to other countries, or tried in U.S. civilian courts. Few countries were interested in taking high-risk prisoners, however; in addition, federal trials of detainees posed enormous procedural and security problems, and the rate of a return to terrorism among released prisoners was high. In May 2009 the administration altered course and announced that it would retain the use of military tribunals, albeit with new procedures that provided additional defendant rights. Also, the one-year deadline for closing the Guantánamo facility was abandoned. Administration officials explored the possibility of confining most of the inmates at an unused state prison in rural Illinois, though opponents in Congress argued that housing the detainees on U.S. soil would imperil national security. At the end of 2010, in an attempt to keep the Guantánamo facility open, Congress attached a rider to a defense bill that purported to forbid transfer of prisoners to the United States for trial and to limit dispersal of terrorist detainees to other countries.

Meanwhile, Obama's attorney general, Eric Holder, declared in November 2009 that Khalid Sheikh Mohammed and four other Guantánamo detainees would stand trial in federal court in New York City on charges stemming from the September 11 attacks. This decision meant that the defendants, all of whom had been captured abroad, would receive most of the constitutional protections and process rights afforded U.S. citizens. Holder defended the venue as appropriate because most September 11 victims were civilians and the attacks occurred on U.S. soil. The plan suffered a major setback in November 2010, however. In the first civilian trial of a former detainee at Guantánamo, a New York City jury acquitted Ahmed Khalfan Ghailani on all but one of 285 counts arising from the 1998 bombings of U.S. embassies in Kenya and Tanzania. The presiding judge had ruled that a key prosecution witness could not testify because the government had learned about him through information obtained from Ghailani at Guantánamo, where the defendant said he had been tortured. The Obama administration found itself facing the politically unacceptable possibility that future trials in federal courts might actually result in the acquittal of a major terrorism suspect. New York officials, meanwhile, continued to object strongly to trying the September 11 coconspirators in Manhattan, and the trial was delayed indefinitely.

PIRACY ON THE HIGH SEAS: INDIAN OCEAN, 2005

To the astonishment of many, high-seas piracy, a crime thought long relegated to legend, made headlines in late 2005 when a luxury cruise ship was attacked off the Somali coast. The *Seabourn Spirit*, carrying 151 Western tourists, managed to evade capture but not without one of its security officers wounded and the ship itself damaged by rocket-propelled grenades. It was a miracle that the ship escaped; since March, 28 vessels had been attacked in the same waters, many of which were hijacked.

In 2005 modern-day piracy was as violent, as costly, and as tragic as it ever had been in the days of yore. Pirates no longer fit the Hollywood image of plundering buccaneers—with eye patches, parrots on their shoulders, cutlasses in their teeth, and wooden legs—but were often ruthless gangs of agile seagoing robbers who attacked ships with assault rifles and antitank missiles. According to the International Maritime Bureau, the organization that investigates maritime fraud and piracy, there had been 325 reported attacks on shipping by pirates worldwide the year before, in 2004. These statistics, the IMB said, reflected only reported incidents directed at commercial shipping and represented a fraction of the actual number. Most acts of piracy went unreported because shipowners did not want to tie up a vessel, costing tens of thousands of dollars a day to operate, for lengthy investigations. The human cost was also high—399 crew members and passengers were killed, were injured, were held hostage, or remained missing at the end of 2004. These statistics did not include, however, those innocent passengers, tourists, commercial fishermen, or yachtsmen whose mysterious disappearances were unofficially attributed to acts of piracy or maritime terrorism.

Piracy Ancient and Modern

Piracy, a crime as old as mankind, has occurred since the earliest hunter-gatherer floated down some wilderness river on a log raft and was robbed of his prized piece of meat. Homer first recorded in *The Odyssey* an act of piracy around 1000 BCE. In many parts of the world, the culture of piracy dates back generations; ransacking passing ships was considered part of local tradition and an acceptable though illegal way of earning a living.

In the modern age, pirates found it relatively easy to attack a ship and make a clean getaway. Sea robbers on small, fast boats could quickly approach the rear of a ship within the blind spot of its radar, toss grappling hooks onto the rail, scale the transom, overpower the crew, and loot the ship's safe. In less than 20 minutes, raiders would be back in their boats, often tens of thousands of dollars richer. Only a few pirates were ever caught, making it clear that plundering a ship was far less risky than robbing a bank.

Historical events and technological innovation also conspired to make modern piracy much easier to commit. Following the end of the Cold War, superpower navies ceased to patrol vital waterways, and local nations were left to deal with problems that heretofore had been international in nature. Pirates no longer had to rely on cotton sails, oars, sextants, and dead reckoning to mount an attack. Modern-day pirates used mobile phones, portable satellite navigation systems, handheld VHF ship-to-ship/shore radios, and mass-produced fibreglass and inflatable dinghies that could accommodate larger and faster inexpensive Japanese outboard motors. Indeed, the pirates who attacked the *Seabourn Spirit* had taken a page from Blackbeard and had launched their attack from a mothership stationed far offshore.

Piracy for Gain or Terror

Several types of piracy existed in 2005. The most common one was the random attack on a passing ship—a mugging at sea. Merchant vessels were slow-moving, lumbering beasts of trade that paraded in a line down narrow shipping lanes. They presented easy targets. The booty for these pirates was crew members' possessions—watches and MP3 players—as well as the cash aboard the ship. A second type of attack was one planned in advance against vessels known to be carrying tens of thousands of dollars in crew payoff and agent fees. With the complicity and connivance of local officials, transnational crime syndicates employed pirates to pillage these vulnerable ships.

Though little known outside the maritime industry, crime syndicates also organized the hijackings of entire ships and cargo. With military precision, a ship carrying cargo that could easily be sold on the black market would be taken over, and it would simply disappear from the face of the Earth; the bodies of the crew would often be found washed up on a deserted shore some days later. The stolen vessel would become a phantom ship, with a new name, new home port, new paint job, and false registration under a different national flag. The vessel would be used to transport drugs, arms, or illegal immigrants or utilized in cargo scams.

Modern piracy took another, even more ominous turn. Pirates discovered that kidnapping the master and another officer was more lucrative than merely stealing the captain's Rolex watch. In 2004 a record 86 seafarers were kidnapped, and in nearly every case a ransom was paid.

There is a long-standing link between piracy and terrorism, and the possibility of terrorism at sea became a growing concern post-September 11, 2001. Maritime terrorism was not new, however. In 1985, members of the Palestine Liberation Front attacked an Italian cruise ship, the *Achille Lauro*, and one of the passengers was shot and thrown overboard; in 2001 Basque separatists attempted to bomb the *Val de Loire* on a passage between Spain and the United Kingdom; and in February 2004 Abu Sayyaf, a terrorist group associated with al-Qaeda, admitted having planted explosives that sank *SuperFerry 14* in Manila Bay. Of the 900 persons aboard that ferry, 116 lost their lives.

Defense

Merchant ships in 2005 had no real defenses against an attack. Fire hoses might blast outboard, decks could be well lighted, and an extra crew member with a handheld radio might patrol the decks, but these precautions were not adequate. They merely indicated to pirates lying in wait that a ship was aware that it had entered pirate territory and that another ship in the vicinity without these obvious defenses might present a softer target. The *Seabourn Spirit* had been a little better equipped than most. She repelled the pirates by use of firehoses as well as a nonlethal acoustic weapon that aimed an earsplitting noise at the attackers; one of the passengers said the pirates fled because they thought the ship was returning fire.

Even the most modern and sophisticated vessel was vulnerable to attack. Suicide bombers in October 2000 had nearly sunk the U.S. destroyer *Cole,* a state-of-the-art warship, and in 2002 suicide terrorists had attacked the modern supertanker *M/V Limburg,* laden with Persian Gulf crude oil in the Gulf of Aden.

Piracy in the Malacca Strait

In 2005 maritime officials were concerned that terrorists might target the world's strategic maritime passages, blocking the movement of global trade. Attention focused on the Malacca Strait, the gateway to Asia, conduit of a third of world commerce, and a prime hunting ground for pirates. About 80 percent of the oil bound for Japan and South Korea was shipped from the Persian Gulf through the strait. In addition, some 50,000 ships transited this narrow channel annually. U.S. officials expressed fears that one day terrorists trained to be pirates—as terrorists trained to be pilots for the September 11 attacks—would take over a high-profile ship and turn it into a floating bomb and close the strait. Disrupting the flow of half the world's supply of oil that is transported through the passage would have a catastrophic effect on the world economy.

Though the U.S. government offered Malaysia and Indonesia (nations through which the strait passes) military patrol boats and personnel to guard the waterway, the offer was quickly rejected by both littoral states on the grounds that the patrolling of their waters by American forces was a violation of territorial sovereignty. Those nations were also mindful that an American military presence in the strait would stir an already restive Muslim population within their countries. By 2005 Malaysia and Indonesia together with the city-state of Singapore, located at the mouth of the strait, had established joint patrols, increased intelligence sharing, and formed a joint radar surveillance project. Issues remained regarding the employment of hot pursuit—one of the most indispensable tools for combating piracy, involving the right to chase pirates back to their lairs in another country's territory.

OUTSOURCING WAR: IRAQ, 2006

The conflict in Iraq focused renewed attention on the role played by private military firms (PMFs) in modern war. In 2006 more than 60 firms employing 20,000 armed personnel were estimated to be operating in Iraq, which made PMFs the second largest foreign military contingent, after the United States. These firms conducted vital security duties, ranging from escorting convoys of freight to protecting key facilities and leaders. The industry even had its own lobby group, the Private Security Company Association of Iraq, with nearly 50 international corporate members. PMFs had also attracted unwanted attention, however, including allegations that contractors working in 2003 as military interrogators and translators at the notorious Abu Ghraib prison near Baghdad were involved in the abuse of prisoners. In March 2006 a jury found the PMF Custer Battles guilty of having defrauded the U.S. government of millions of dollars for work done while under contract in Iraq.

The Evolution of PMFs

The term PMF—also private security company and military services provider—was a catch-all expression that included traditional security firms employing armed guards, companies shipping defense matériel, consultants offering advice on strategy, and military trainers. Unlike traditional defense industries, PMFs operated in combat zones and other areas where violence might be imminent. States, private industry, and humanitarian aid agencies all employed the services of PMFs.

The modern PMF was a product of the end of the Cold War; in the early 1990s many countries slashed defense budgets following the demise of the Soviet Union. This coincided with the growing trend of governments to outsource services to private industry. As a consequence, armed forces were left to carry out their missions with fewer ships, aircraft, and personnel, leaving more support and rear-area functions (e.g., repairing tanks, training pilots, and preparing meals) to be outsourced to contractors.

It would be wrong, however, to conclude that PMFs were newcomers to warfare. Prior to the 19th century, it was common for states to contract for military services, including combat. The word *soldier* itself is derived from the Latin *solidus,* meaning a gold coin. During the 3rd century BCE, Alexander I the Great employed mercenary forces to help conquer Asia, and during the American Revolution (1775–83) Britain hired German soldiers called Hessians to fight the colonists. In the 17th and 18th centuries, the British East India Company and its Danish, Dutch, and French rivals all had private armies to help defend their government-sanctioned business interests in Asia.

Effects on the Military

The growth of the modern privatized military industry had an effect on the armed forces that they were intended to assist. With PMFs offering daily wages of up to $1,000 to attract highly trained staff, there was an exodus of soldiers from many special forces. Britain's Special Air Service, the U.S. Army's Special Forces, and the Canadian Army's Joint Task Force 2 all acknowledged problems retaining personnel and were offering special bonuses and pay increases in an effort to compete with lucrative wages in the private sector.

When a military organization has no organic capability, it becomes dependent on private industry to provide it. In 2000, for example, the Canadian navy had no logistics ships, and the government contracted a shipping company to take 580 vehicles and 390 sea containers full of equipment back to Canada following the completion of NATO operations in Kosovo. Owing to a dispute over unpaid bills, the ship loitered in international waters for two weeks until Canadian military personnel boarded the ship and forced it to dock in a Canadian port.

Despite such problems, PMFs found themselves called upon to deliver services previously considered the domain of military personnel. Kellogg, Brown & Root (KBR), for instance, ran the only permanent U.S. base in Africa (Camp Lemonier in Djibouti, at the mouth of the Red Sea). KBR had more than 700 employees who did laundry, cleaned buildings, and prepared meals for 1,500 military personnel. PMFs had even been employed by governments to handle domestic emergencies, such as the initial response to Hurricane Katrina in New Orleans in 2005.

After the Sept. 11, 2001, attacks in the United States, the war on terrorism declared by the U.S. government provided new opportunities for PMFs. Spy agencies turned to PMFs to collect and analyze intelligence from around the world. At times, contractors outnumbered employees at the CIA's offices in both Iraq and Pakistan.

Legal Issues

International humanitarian law (which includes the Geneva Conventions) applied to every person in a war zone, even though the status of PMFs was not specifically defined. Hence, PMF employees were considered civilians and could not be targeted for attack unless they formed part of the armed forces of a state. If these employees participated directly in hostilities, however, they would lose this legal protection. Furthermore, PMF employees participating directly in hostilities were not entitled to protection as prisoners of war under the Geneva Conventions, and they could be tried as "unlawful combatants" (in other words, as mercenaries). The distinction between combatants and civilians who were merely defending themselves became complicated when PMF staff wore military clothing and carried government-issued or privately owned weapons. According to the International Committee of the Red Cross, if a state outsourced military functions to a PMF, the state would remain legally responsible for the firm's acts.

Another legal problem was that PMF employees were usually exempt from the military laws that governed how troops behave in a conflict. For example, although by 2006 soldiers from several coalition members had been convicted of crimes against civilians since the U.S.-led invasion of Iraq in March 2003, not a single military contractor had been charged with a crime there.

Although most states published statistics on the numbers of their military casualties, the fate of PMF personnel went largely unreported in the news media. With few exceptions—such as the horrific public display of murdered contractors in the Iraqi city of Fallujah in March 2004—there was little news coverage of the nearly 650 civilian contractors working for the U.S. government who were reportedly killed in Iraq between March 2003 and September 2006. Safety became another area of concern, especially when the responsibility for the safety of PMF employees working in war zones was undefined.

Although some countries prohibited their citizens from joining the armed forces of a foreign country at war, very few prevented them from joining foreign PMFs. In 2006 the South African parliament introduced legislation to prevent any of its citizens from participating in a foreign conflict. The bill—though it languished in parliament—had its genesis in a 2004 coup attempt against the president of Equatorial Guinea. Mark Thatcher, the son of former British prime minister Margaret Thatcher and a resident of South Africa at the time, helped fund the PMF hired to conduct the coup, and it in turn hired 70 South Africans to do the fighting.

Globally, the use of PMFs had grown dramatically since the 1990–91 Persian Gulf War, when there was an estimated one contractor for every 50 military personnel involved. By the time of the Iraq invasion in 2003, the ratio had grown to one in 10. By 2006, with PMFs operating on nearly every continent and generating an estimated $100 billion in revenue annually, they were certain to remain important actors in modern warfare for the foreseeable future.

ADVANCES IN BATTLEFIELD MEDICINE: IRAQ AND AFGHANISTAN, 2007

In the wars being fought in Afghanistan and Iraq, better battlefield medical care, together with the use of advanced body armour and helmets, was leading to survival rates higher than had been seen in previous wars. Through mid-2007, about 40 percent of coalition casualties in Iraq had been caused by improvised explosive devices, such as explosive vests used by suicide bombers, hand grenades rigged with trip wires, and sophisticated roadside bombs detonated by remote control. Another 30 percent of casualties had been caused by gunfire. The remaining 30 percent of casualties had various causes such as mortar attacks, vehicle crashes, and stabbings.

Studies of historical casualty rates had shown that about one-half of military personnel killed in action died from the loss of blood and that up to 80 percent of those died within the first hour of injury on the battlefield. This time period had been dubbed the "golden hour," a brief interval when prompt treatment of bleeding had the best chance of preventing death. Modern developments in military medicine therefore focused on treatment to stop bleeding quickly and on the provision of prompt medical care.

Battlefield First Aid

All modern troops were trained in the basics of first aid, including how to stop bleeding, splint fractures, dress wounds and burns, and administer pain medication. Combat troops were issued a first-aid kit that included a tourniquet that could be applied with one hand. (Though the use of tourniquets had previously been considered undesirable, the military now regarded them as lifesaving tools for severe limb wounds.) Also, new pressure dressings had been issued that could clot severe bleeding within seconds of being applied. These dressings include HemCon, made with chitosan (an extract from shrimp shells), and QuikClot, made with inorganic zeolite granules.

Within every military unit, there were personnel specially trained to provide medical assistance to the wounded in order to stabilize their condition until they could be treated by a physician. For example, a typical U.S. Army battalion of 650–690 combat soldiers had 20–30 such medics (called corpsmen in the U.S. Marines), who were trained in the identification and assessment of different types of wounds as well as in advanced first aid, such as administering intravenous fluids and inserting breathing tubes. Modern medic training made use of sophisticated lifelike mannequins programmed to simulate various injuries and to respond to treatment. Some training could also involve the use of mammals anesthetized under the supervision of veterinarians so that the medic gained experience with real injuries on live tissue.

Evacuation for Treatment and Surgery

As soon as the situation permitted, the wounded would be taken from the scene of the battle to their unit's closest treatment facility, which served as a collection point for casualties and was maintained as close to the battlefield as possible. The facility, which might be a battalion aid station or regimental aid post, was staffed by one or more physicians whose task was to stabilize patients further and to assess them for transfer to better-equipped facilities. The rapid evacuation of wounded personnel to medical facilities for higher-level care was crucial to saving lives within the "golden hour." Helicopters provided the most important means of medical evacuation. The HH-60M Blackhawk helicopter used by the U.S. Army had environmental-control and oxygen-generating systems, patient monitors, and an external rescue hoist. In 2005 the U.S. Army began deploying to Iraq a new variant of the eight-wheeled Stryker armoured vehicle to be used as a medical evacuation vehicle. It was faster and better protected than previous military ambulances, and it could carry up to six patients while its crew of three medics provided medical care.

The mobile army surgical hospital (MASH) was used by U.S. forces during the Korean War in the 1950s and was still in service during the Persian Gulf War (1990–91). MASH units—which had 60 beds, required 50 large trucks to move, and took 24 hours to set up—were deemed too cumbersome to keep up with fast-moving armoured and airmobile forces, and they were supplanted by the smaller Forward Surgical Team (FST). The FST comprised 20 persons, including 4 surgeons, and it typically had 2 operating tables and 10 litters set up in self-inflating shelters. It could be deployed close to the battlefield and made operational in one and a half hours. FSTs were designed not to hold patients for any length of time but to stabilize them enough to be transported to a

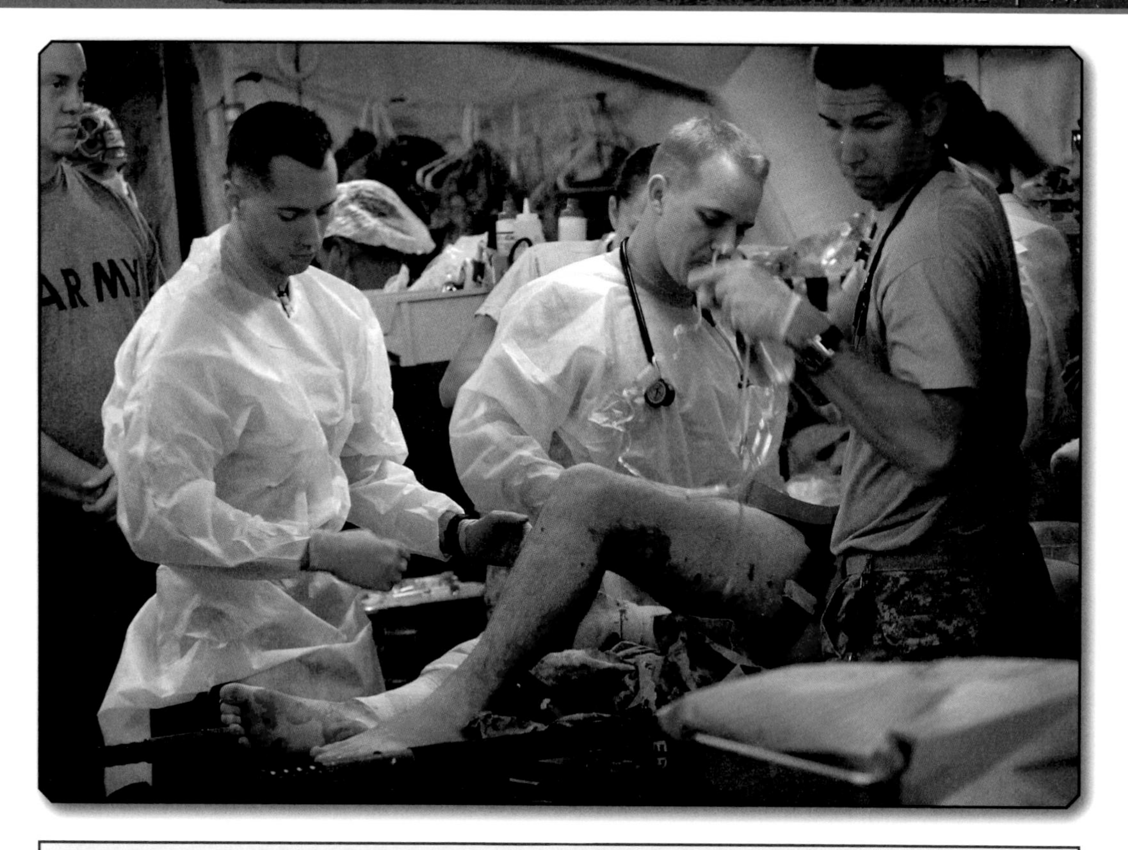

A team at the 31st Combat Support Hospital working on a wounded marine at Camp Dwyer, Sept. 24, 2010, near Marja, Afghan. Scott Olson/Getty Images

larger facility with more specialized staff and equipment.

For most U.S. casualties in Iraq and Afghanistan, the first fully equipped surgical facility they reached was the Combat Support Hospital (CSH).

The CSH staff included specialists such as orthopedic and oral surgeons and psychiatrists. The CSH was modular in design and could be configured in sizes from 44 to 248 beds as needed. It was assembled from metal shelters and climate-controlled tents, complete with water and electricity. The facility had an intensive-care unit, operating theatres, a radiography section (with X-ray machine and CT scanner), pharmacy, and laboratory for banking whole blood. Although the use of fresh whole-blood transfusions declined in civilian hospitals after the 1950s, it was still used to treat combat casualties because it retained its ability to clot far better than frozen stored blood. In 2004 military doctors began using an experimental blood-clotting drug called recombinant activated factor VII to treat severe bleeding, despite some medical evidence that linked it to deadly blood clots.

Military medicine had also benefited from advances in digital technology. For example, military hospitals in Afghanistan and Iraq had CT scanners and ultrasound machines with Internet links to medical specialists to allow military doctors to consult with the specialists about detailed diagnosis and treatment. Also, patients could have their medical records transmitted electronically to any hospital to which they were transferred for further treatment.

Once treated at a CSH, the most serious American casualties in Iraq and Afghanistan were transported by fixed-wing aircraft to Landstuhl Regional Medical Center in Germany, a 10- to 12-hour flight. For even higher-level care, they were then transported to military hospitals in the United States.

Extended Care and Rehabilitation

Wounded personnel who could not be returned to duty received extended care and rehabilitation. Modern body armour and helmets had reduced the incidence of lethal penetrating wounds to the torso and head, but as a result more wounded soldiers had been surviving with debilitating injuries, such as the loss of one or more limbs. By 2007 the number of wounded U.S. veterans who had returned from the wars in Afghanistan and Iraq was straining the military health care system. Notably, the best-known U.S. military hospital, Walter Reed Army Medical Center in Washington, D.C., came under scrutiny early in the year over substandard outpatient treatment, and a presidential commission was established to examine the quality of health care provided to veterans.

One of the challenges facing military medicine was the treatment of post-traumatic stress disorder and other psychological damage resulting from service in a war zone. According to some studies available in 2007, up to one-third of U.S. soldiers returning from Afghanistan and Iraq had at least one mental health problem. With more soldiers surviving the loss of an arm or a leg, there was also the challenge of developing better prosthetics. One example was the bionic hand called i-Limb, which became available to amputees in 2007. The prosthetic had five fully and independently functional fingers and was controlled by a computer chip connected to electrodes that detected electrical signals from surviving arm muscles.

CHILD SOLDIERS: AFRICA AND ASIA, 2009

In April 2009, 112 child soldiers who had served with the rebel National Liberation Forces (FLN) were freed following the signing of a cease-fire agreement between the FLN and the government of Burundi, and at that moment the existence of modern-day child soldiers was brought forcefully into the international spotlight. Worldwide, armed forces and nongovernmental armed groups were recruiting and exploiting children, who were defined under international law as those under 18 years of age. Though the number of child soldiers was unknown—many child recruiters were successfully hiding their actions, and some children were lying about their age in order to join political struggles—it was estimated that at any one time there were approximately 250,000 child soldiers, many of them girls. Although most child soldiers were teenagers, the recruits also included children as young as six or seven years of age. Children might also be born into armed groups. For example, the Lord's Resistance Army (LRA), which had abducted many children and had been fighting the government of Uganda since the 1980s, maintained military camps in southern Sudan, where its leader, Joseph Kony, sired many children who subsequently became soldiers.

Armed forces and groups recruited children for diverse reasons. Commanders often selected children because they were available in large numbers and could be recruited with impunity, because they could be fashioned into effective fighters, and because commanders knew that

Former child soldiers leaving a ceremony that turned them over to the United Nations on June 4, 2001, in Kaliahun, Sierra Leone. Chris Hondros/Getty Images

they could manipulate children easily by employing terror tactics and offering incentives for bravery and initiative in combat. Armed with small lightweight weapons, such as AK-47 assault rifles, even young children could be effective fighters. They might also serve as spies who could slip behind enemy lines without suspicion. Teenagers were often sought for their size and strength, their willingness to take risks that many adults would avoid, and their political consciousness. In Sri Lanka the Liberation Tigers of Tamil Elam fought government forces in part by recruiting teenage girls to serve as suicide bombers.

The Recruitment of Children

Child recruitment was contextual and might involve force or decisions made by the child. The LRA forcibly recruited as many as 60,000 children by abducting and subjugating them into obedience through a regime of terror. To deter escape the LRA forced abducted children to surround recaptured escapees and beat them to death. Forced recruitment was also used in Sierra Leone, where the opposition group Revolutionary United Front forced young people at gunpoint to join and often required children to kill members of their own villages or families.

Some children decided to join armed groups, but their choices might not be "voluntary," since they were made in desperate circumstances and involved a mixture of "push" and "pull" factors. In Colombia, for example, a boy who had been abused in his home might leave and seek an alternate "family" in the form of an armed group. In other countries youths had been lured by propaganda and an ideology of liberation into believing that by becoming soldiers, they would help to liberate their people. In Rwanda young Hutu were recruited into a youth militia (the Interahamwe) and were taught that Tutsi had to be eliminated; more than 800,000 people, mostly Tutsi, were killed in the 1994 genocide.

Other pull factors might include retribution, money, family ties, and power. In Liberia some children joined armed groups in an effort to avenge wrongs, such as the killing of one's parents by government forces. Children might also be eager to earn money that they could send home to support impoverished families. In northern Afghanistan children frequently joined the Northern Alliance to fight the Taliban because their fathers, brothers, or uncles were members and because they regarded fighting as a matter of family honour and village protection. Some children sought power and prestige. Many children reported that because they carried a gun and wore a uniform, they were treated with a level of respect that they never enjoyed as civilians.

Inside armed groups, children played diverse roles. A common myth stated that all child soldiers were fighters, when in fact many recruits served as porters, cooks, bodyguards, and domestics, among other roles. Another myth was that all child soldiers were boys. In conflicts in countries such as Liberia and Sierra Leone, girls were recruited to serve not only as fighters but also as sex slaves, whose refusal to provide sex often led to severe punishment or death.

The Reintegration of Child Soldiers

Because they had been socialized into lives as soldiers, child soldiers might themselves become a means of perpetuating violence and armed conflict. To break cycles of violence, it became a key priority to demobilize child soldiers and help them to transition or reintegrate into civilian life. Typically, this was being done through a process of disarmament, demobilization, and reintegration. Having turned in their weapons (disarmament), child soldiers were demobilized by being officially stood down from armed groups. They were then reintegrated through rehabilitation and work with families and communities to help them find a place in civilian society.

Rehabilitation required attention to mental health issues that might cause distress and impede reintegration. In countries such as Liberia and Sierra Leone, where commanders had plied child soldiers with drugs to make them fearless, many former child soldiers developed problems of substance abuse. In other countries a minority of former child soldiers developed clinical problems, such as depression, anxiety, and trauma, particularly the post-traumatic stress disorder that can arise following extreme events, including exposure to deaths or active engagement in killing. Effective treatment of these problems required specialized supports, such as counseling by well-trained psychologists or psychiatrists, few of whom were available in war zones. In addition, mental health issues might have indigenous roots. In Angola, for example, former child soldiers were terrified because they believed that they were haunted by the unavenged spirits of the people they had killed. In this case, rather than counseling, the children would benefit from the services of a traditional healer, who might conduct a cleansing ritual to remove their spiritual pollution.

It was often everyday social issues, however, that caused the greatest distress and the most formidable barriers to reintegration. To rectify family separation it became essential, when possible, to reunify former recruits with their families and to manage family conflicts. Nearly all former child soldiers struggled because they had lost years of education and lacked the income needed to start a family or the social skills to assume the role of mother or father. Some developed unruly behaviour, while others had difficulty meeting expectations associated with ordinary living. Many former child soldiers—particularly girls—were stigmatized and called "rebels" or were viewed as aggressive troublemakers. Media accounts sometimes supported these stereotypes by referring to former child soldiers as a "Lost Generation."

Effective reintegration was being made possible through holistic community-based supports. It became important to mobilize communities to support the livelihood, acceptance, and education of former child soldiers and to activate protection mechanisms that guarded against rerecruitment or retaliation. Nevertheless, reintegration efforts were not sufficient by themselves; equal efforts had to be given to prevention, particularly to ending the impunity that allowed recruitment to continue.

Recognition by the international community of the serious nature of enlisting children in warfare was highlighted in 2009 when warlord Thomas Lubanga Dyilo became the first person to be tried by the International Criminal Court. He was accused of having committed war crimes (recruiting children as soldiers in the Democratic Republic of the Congo). The UN was also at the forefront of strengthening international standards against child recruitment and urging governments to ratify the Optional Protocol on the involvement of children in armed conflict. The optional protocol, which went into effect in 2002 to augment the UN Convention on the Rights of the Child (1989), would raise the minimum age of participation in hostilities from 15 years of age to 18 for any country that ratified it. By 2009 some 130 countries had done so, but the effort would truly succeed only if all countries agreed to abide by the protocol and thus safeguard the world's children.

UAVS: MIDDLE EAST AND SOUTH ASIA, 2009

A little-known but important milestone in modern warfare was reached in 2009: in that year the U.S. Air Force trained more operators of unmanned aerial vehicles (UAVs) than it did pilots. In an age when war was increasingly dominated by robots, the U.S. military alone fielded at least 7,000 of these machines, which were either remotely guided by a human using a radio link or self-guided by preprogrammed flight plans. Interest in UAVs was global, however. More than 60 manufacturers in at least 40 countries were now servicing a market that was expected to exceed tens of billions of dollars over the following decade. It was not surprising, then, that Quentin Davies, the U.K.'s minister for defense equipment and support, predicted in July 2009 that the world was witnessing the last generation of manned combat aircraft and that by 2030 UAVs would displace them.

A Growing Technology

UAVs, also called remotely piloted vehicles (RPVs) or unmanned aircraft systems (UASs), were aircraft without a pilot onboard. Fixed-wing UAVs resembled "smart weapons" such as cruise missiles, but they were superior because they returned to their base after a mission and could be reused. Also, UAVs had two decisive advantages over manned aircraft: their use did not risk the lives of aircrews, and they could loiter over areas of interest longer than most types of aircraft with human pilots. The current generation of UAVs varied in size from small propeller-driven hand-launched models such as the German army's Aladin to jet-powered intercontinental-range craft such as the U.S. Air Force's RQ-4 Global Hawk. Prices

A U.S. marine inspecting an unmanned aerial vehicle (UAV) in Camp Dwyer in Gamser, Helmand Province, Afghan., on Feb. 15, 2011. Adek Berry/AFP/Getty Images

ranged from a few hundred thousand dollars for small models to well over $100 million for a Global Hawk.

UAVs first took to the skies during World War II with radio-controlled target drones, and they continued to develop slowly through the Vietnam era, when film cameras were mounted onto jet-powered drones for photoreconnaissance missions. Truly modern UAVs did not begin to appear over battlefields until the 1980s, when a number of technical advancements made them much more effective. Advanced composite materials made for lighter, stronger airframes, and improved electronics permitted the development of high-resolution television and infrared cameras. Also, full implementation of the Global Positioning System (GPS) in the 1990s made it possible to navigate UAVs with a precision that was previously unattainable.

UAVs began to garner media attention during NATO's intervention in the Yugoslav civil war of the 1990s. In 1995 the U.S. Air Force put the RQ-1 Predator into service for airborne surveillance and target acquisition. With its pusher propeller driven by a four-cylinder gasoline engine, the Predator could cruise at 140 km (87 miles) per hour, stay aloft for up to 16 hours, and reach altitudes of 7,600 metres (25,000 feet). Predators flying over Yugoslavia tracked troop movements, monitored refugees, and marked targets so that manned aircraft could attack them with laser-guided bombs.

The Predator remained the most widely used battlefield UAV operated by the United States into 2009. The entire system consisted of the vehicle itself (with built-in radar, TV and infrared cameras, and laser designator), a ground-control station, and a communication suite to link the two by satellite. Though pilotless, the Predator was operated by approximately 55 personnel, including a pilot operator and a sensor operator as well as intelligence, maintenance, and launch and recovery specialists. The latest version, designated the MQ-1, had gone into service in 2001 armed with two laser-guided AGM-114 Hellfire missiles, giving the UAV the ability to attack targets as well as identify them. The most celebrated Predator kill was in Yemen in 2002, when a craft operated by the CIA destroyed a vehicle carrying six members of al-Qaeda. A turboprop-powered version of the Predator, called the MQ-9 Reaper, was significantly larger and had a greater payload. The Reaper had been operational since 2007 with U.S. forces and was also used by Britain's Royal Air Force.

A Mixed Record

UAV technology might be sophisticated, but it was still in its infancy. By 2009 some 65 Predators (each costing $4 million) had crashed, including at least 3 in 2009. Thirty-six of the crashes were attributed to human error. Since UAVs were not yet completely autonomous, their operators had to display great skill in judging distance and speed when landing, a task made more difficult by a slight delay in signal transmission between the UAV and the ground-control station. Moreover, there were occasional technical glitches, such as one that occurred in September 2009 when a Reaper on a combat mission over Afghanistan could no longer be controlled and had to be shot down by U.S. warplanes.

Successful attacks by UAVs depended upon the accuracy and timeliness of intelligence. This principle was demonstrated in 2009 when an al-Qaeda military planner was believed to have been killed by an American UAV in September but suddenly appeared in a media interview in October. Also, high-tech weapons might win engagements on the battlefield, but they could not solve political problems—and on occasion they may even have aggravated them. In July 2009 the Brookings Institution think tank estimated that for every militant killed by a UAV in Afghanistan and Pakistan, approximately 10 civilians were also killed, a situation that was alienating the local population and turning them against the United States and its NATO allies. UAV use also raised issues of accountability. According to the nongovernmental organization Human Rights Watch, Israeli UAVs unlawfully killed at least 29 Palestinian civilians during the Gaza incursion in late 2008 and early 2009 because UAV operators allegedly failed to verify that targets were combatants.

Future Uses in War and Peace

Most UAVs remained dedicated to what the military called ISTAR—intelligence, surveillance, target acquisition, and reconnaissance. For example, American UAVs began patrolling off the coast of Somalia in October 2009 in order to provide early warning of pirate vessels approaching merchant ships and to guide naval forces. However, the number of potential uses for UAVs was growing. In August 2009 the U.S. Marine Corps awarded contracts to Boeing and a joint venture between Lockheed Martin and Kaman to develop cargo UAVs that would be capable of delivering supplies to troops on the battlefield. The goal was to demonstrate how such UAVs could reduce risk and expense in logistics. It was expensive to operate ground supply convoys on the poor roads and in the back country of Afghanistan; also, convoys had to be heavily guarded, and they continually ran risks from roadside bombs and ambushes.

Besides these military uses, UAV technology was attracting interest from police forces and other civilian agencies. For example, the U.S. Customs and Border Patrol had been using the Predator to patrol the Mexico-U.S. border since 2005 and the Canada-U.S. border since early 2009, and two maritime-patrol variants were scheduled to be operational in 2010. UAVs were also being developed for use in search-and-rescue operations to help locate survivors and deliver emergency supplies to them. In addition, UAVs were being evaluated for their potential in assessing damage suffered from disasters such as hurricanes, forest fires, and maritime oil spills.

As robotic vehicles became more commonplace, UAVs could be expected to be used wherever possible to minimize threats to personnel and to do tasks that exceeded human strength and endurance. If the trends of 2009 continued, UAVs could one day evacuate casualties from the heat of battle and mount round-the-clock surveillance missions for months and maybe even years at a time.

WAR ON COMPUTERS AND NETWORKS: CYBERSPACE, 2010

Computers and the networks that connect them are collectively known as the domain of cyberspace, and in 2010 the issue of security in cyberspace came to the fore, particularly the growing fear of cyberwar waged by other states or their proxies against government and military networks in order to disrupt, destroy, or deny their use. In the United States, Sec. of Defense Robert Gates on May 21 formally announced the appointment of Army

U.S. Air Force personnel updating antivirus software for protection against cyberspace hackers, Barksdale Air Force Base, La., 2010. Tech. Sgt. Cecilio Ricardo/U.S. Air Force

Gen. Keith B. Alexander, director of the National Security Agency (NSA), as the first commander of the newly established U.S. Cyber Command (USCYBERCOM). The announcement was the culmination of more than a year of preparation by the Department of Defense. Soon after a government Cyberspace Policy Review was published in May 2009, Gates had issued a memorandum calling for the establishment of USCYBERCOM, and Alexander underwent months of U.S. Senate hearings before he was promoted to a four-star general in May 2010 and confirmed in his new position. USCYBERCOM, based at Fort Meade, Md., was charged with conducting all U.S. military cyberoperations across thousands of computer networks and with mounting offensive strikes in cyberspace if required.

Attacks in Cyberspace

Western states depended on cyberspace for the everyday functioning of nearly all aspects of modern society, and developing states were becoming more reliant upon cyberspace every year. Everything modern society needed to function—from critical infrastructures and financial institutions to modes of commerce and tools for national security—depended to some extent upon cyberspace. Therefore, the threat of cyberwar and its purported effects were a source of great concern for governments and militaries around the world, and several serious cyberattacks had taken place that, while not necessarily meeting a strict definition of cyberwar, could serve as an illustration of what might be expected in a real cyberwar of the future.

The cyberspace domain was composed of three layers. The first was the physical layer, including hardware, cables, satellites, and other equipment. Without this physical layer, the other layers could not function. The second was the syntactic layer, which included the software providing the operating instructions for the physical equipment. The third was the semantic layer and involved human interaction with the information generated by computers and the way that information was perceived and interpreted by its user. All three layers were vulnerable to attack. Cyberwar attacks could be made against the physical infrastructure of cyberspace by using traditional weapons and combat methods. For example, computers could be physically destroyed, their networks could be interfered with or destroyed, and the human users of this physical infrastructure could be suborned, duped, or killed in order to gain physical access to a network or computer. Physical attacks usually occurred during conventional conflicts, such as in the North Atlantic Treaty Organization's (NATO's) Operation Allied Force against Yugoslavia in 1999 and in the U.S.-led operation against Iraq in 2003, where communication networks, computer facilities, and telecommunications were damaged or destroyed.

Attacks could be made against the syntactic layer by using cyberweapons that destroyed, interfered with, corrupted, monitored, or otherwise damaged the software operating the computer systems. Such weapons included malware, malicious software such as viruses, trojans, spyware, and worms that could introduce corrupted code into existing software, causing a computer to perform actions or processes unintended by its operator. Other cyberweapons included distributed denial-of-service, or DDoS, attacks, in which attackers, using malware, hijacked a large number of computers to create so-called botnets, groups of "zombie" computers that then attacked other targeted computers, preventing their proper function. This method was used in cyberattacks against Estonia in April and May 2007 and against Georgia in August 2008. On both occasions it was alleged that Russian hackers, mostly civilians, conducted denial-of-service attacks against key government, financial, media, and commercial Web sites in both countries. These attacks temporarily denied access by the governments and citizens of those countries to key sources of information and to internal and external communications.

Finally, semantic cyberattacks, also known as social engineering, manipulated human users' perceptions and interpretations of computer-generated data in order to obtain valuable information (such as passwords, financial details, and classified government information) from the users through fraudulent means. Social-engineering techniques included phishing—in which attackers sent seemingly innocuous e-mails to targeted users, inviting them to divulge protected information for apparently legitimate purposes—and baiting, in which malware-infected software was left in a public place in the hope that a target user would find and install it, thus compromising the entire computer system. In August 2010, for example, fans of the Anglo-Indian movie star Katrina Kaif were lured into accessing a Web site that was supposed to have a revealing photograph of the actress. Once in the site, visitors were automatically forwarded to a well-known social-networking site and asked to enter their login and password. With this information revealed by users, the phishing expedition was successfully completed. An example of baiting involved an incident in 2008 in which a flash memory drive infected with malware was inserted into the USB port of a computer at a U.S. military base in the Middle East. From there the computer code spread through a number of military networks, preparing to transfer data to an unnamed foreign intelligence service, before it was detected. As these above examples suggest, semantic methods were used mostly to conduct espionage and criminal activity.

Cybercrime, Cyberespionage, or Cyberwar?

One of the first references to the term *cyberwar* could be found in "Cyberwar Is Coming!," a landmark article by John Arquilla and David Ronfeldt, two researchers for the RAND Corporation, published in 1993 in the journal *Comparative Strategy*. The term was becoming increasingly controversial by 2010, however. A number of experts in the fields of computer security and international politics questioned whether the term accurately characterized the hostile activity occurring in cyberspace. Many suggested that the activities in question could be more accurately described as crime, espionage, or even terrorism but not necessarily as war, since the latter term had important political, legal, and military implications.

For example, it was far from apparent that an act of espionage by one state against another via cyberspace would equal an act of war—just as traditional methods of espionage had rarely, if ever, led to war. Allegations of Chinese cyberespionage bore this out. A number of countries, including India, Germany, and the United States, believed that they had been victims of Chinese cyberespionage efforts. Nevertheless, while these incidents had been a cause of tension between China and the other countries, they had not damaged overall diplomatic relations. Similarly, criminal acts perpetrated in and from cyberspace by individuals or groups were viewed as a matter for law enforcement rather than the military, though there was evidence to suggest that Russian organized-crime syndicates helped to facilitate the cyberattacks against Georgia in 2008 and that they were hired by either Hamas or Hezbollah to attack Israeli Web sites in January 2009. On the other hand, a cyberattack made by one state against another state, resulting in damage against critical infrastructures such as the electrical grid, air traffic control systems, or financial networks, might legitimately be considered an armed attack if attribution could be proved. An example here would be the Stuxnet worm, which in 2010 appeared to have been targeted at machinery used by Iran in its nuclear-energy program (which many states believed was working toward building nuclear weapons).

Some experts specializing in the laws of armed conflict questioned the notion that hostile cyberactivities could cause war (though they were more certain about the use of hostile cyberactivities during war). They argued that such activities and techniques did not constitute a new kind of warfare but simply were used as a prelude to, and in conjunction with, traditional methods of warfare. Indeed, in recent years cyberwar had assumed a prominent role in armed conflicts, ranging from the Israeli-Hezbollah conflict in Lebanon in 2006 to the Russian invasion of Georgia in 2008. In these cases cyberattacks had been launched by all belligerents before the actual armed conflicts began, and cyberattacks had continued long after the shooting stopped, yet it could not be claimed that the cyberattacks launched before the start of actual hostilities had caused the conflicts. Similarly, the cyberattacks against Estonia in 2007 had been conducted in the context of a wider political crisis surrounding the removal of a Soviet war memorial from the city centre of Tallinn to its suburbs, causing controversy among ethnic Russians in Estonia and in Russia itself.

Such qualifications aside, it was widely believed that cyberwar not only would feature prominently in all future conflicts but would probably even constitute the opening phases of them. The role and prominence of cyberwar in conventional conflicts was continuing to escalate.

Cyberattack and Cyberdefense

Despite its increasing prominence, there remained many challenges for both attackers and defenders engaging in cyberwar. Cyberattackers had to overcome cyberdefenses, and both sides had to contend with a rapid offense-defense cycle. Nevertheless, the offense dominated in cyberspace because any defense had to contend with attacks on large networks that were inherently vulnerable and run by fallible human users. In order to be effective in a cyberattack, the perpetrator had to succeed only once, whereas the defender had to be successful over and over again.

Another challenge of cyberwar was the difficulty of distinguishing between lawful combatants and civilian noncombatants. One of the significant characteristics of cyberspace was the low cost of entry for anyone wishing to use it. As a result, it could be employed by anyone who could master its tools. The implications of this openness for cyberwar were that civilians, equipped with the appropriate software, were capable of mounting and participating in cyberattacks against state agencies, nongovernmental organizations, and individual targets. The legal status of such individuals, under the laws of armed conflict and the Geneva Conventions, was unclear, presenting additional difficulty for those prosecuting and defending against cyberwar. The cyberattacks against Estonia and Georgia were examples of this challenge: it was alleged that most, if not all, of those participating in the attacks were civilians perhaps motivated by nationalist fervour.

Perhaps the greatest challenge for states defending against cyberattacks was the anonymity of cyberspace. Mention is made above of the low cost of entry into cyberspace; another major attribute was the ease with which anyone using the right tools could mask his identity, location, and motive. For example, there was little solid evidence linking the Russian government to the Estonia and Georgia cyberattacks, and so one could only speculate as to what motivated the attackers if they did not act directly on orders from Moscow. Such easy anonymity had profound implications for states or agencies seeking to respond to—and deter—cyberwar attacks. If the identity, location, and motivation of an attack could not be established, it would become very difficult to deter such an attack, and using offensive cybercapabilities in retaliation would carry a strong and often unacceptable risk that the wrong target would face reprisal.

Despite these challenges, defending against cyberwar had become a priority for many nations and their militaries. Key features of any major cyberdefense structure included firewalls to filter network traffic, encryption of data, tools to prevent and detect network intruders, physical security of equipment and facilities, and training and monitoring of network users. A growing number of modern militaries were creating units specifically designed to defend against the escalating threat of cyberwar, including the U.S. Air Force and the U.S. Navy, both of which formed new commands under USCYBERCOM. In the United Kingdom the Government Communications Headquarters (GCHQ) created a Cyber Security Operations Centre (CSOC) in September 2009, and France set up its Network and Information Security Agency in July 2009.

Finally, while the public focus of 2010 was on defending against cyberattacks, the use of offensive cybercapabilities was also being considered. There were legal, ethical, and operational implications in the use of such capabilities stemming from many of the challenges mentioned above. Hence, in many Western countries such capabilities were proscribed extensively by law and were alleged to be the preserve of intelligence agencies

Stuxnet

Stuxnet was a computer worm discovered in June 2010 that had been specifically written to take over certain programmable industrial control systems and cause the equipment run by these systems to malfunction, all the while feeding false data to the systems' monitors indicating that the equipment was running as intended.

As analyzed by computer security experts around the world, Stuxnet targeted certain "supervisory control and data acquisition" (SCADA) systems manufactured by the German electrical company Siemens AG that control machinery employed in power plants and similar installations. More specifically, the worm targeted only Siemens SCADA systems that were used in conjunction with frequency-converter drives, devices that control the speed of industrial motors, and even then only drives that were made by certain manufacturers in Finland and Iran and were programmed to run motors at very specific high speeds. This combination indicated to analysts that the likely target of Stuxnet was nuclear installations in Iran—either a uranium-enrichment plant at Naṭanz or a nuclear reactor at Būshehr or both—a conclusion supported by data showing that, of the approximately 100,000 computers infected by Stuxnet by the end of 2010, more than 60 percent were located in Iran.

The worm was found to have been circulating since at least mid-2009, and indeed in the latter part of that year at the Naṭanz plant an unusually large number of centrifuges (machines that concentrate uranium by spinning at very high speeds) were taken out of operation and replaced. The Iranian nuclear program, which most foreign governments believed was working to produce nuclear weapons, continued to suffer technical difficulties even after the discovery of the worm. Though it was impossible to verify that these difficulties were caused by the Stuxnet worm, it became clear to cybersecurity experts that Iran had suffered an attack by what may have been the most sophisticated piece of malware ever written. By taking over and disrupting industrial processes in a significant sector of a sovereign state, Stuxnet may have been the first truly offensive cyberweapon, a significant escalation in the growing capability and willingness of states to engage in cyberwar.

Speculation then centred on where the worm may have originated. Many analysts pointed to the United States and Israel as two countries whose assessments of the threat of Iranian nuclear weapons had long been particularly severe, and whose expertise in engineering and computer science would certainly have enabled them to plan and launch such a cyberattack. Officials of both countries refused to discuss the issue; meanwhile, the Iranian government declared that a foreign virus had infected computers at certain nuclear facilities but had caused only minor problems.

such as the National Security Agency (NSA) in the United States and GCHQ in the United Kingdom. In China, where the legal, ethical, and operational implications differed (or at least appeared to), it was believed that organizations such as the General Staff Department Third and Fourth Departments, at least six Technical Reconnaissance Bureaus, and a number of People's Liberation Army (PLA) Information Warfare Militia Units were all charged with cyberdefense, attack, and espionage. Similarly, it was thought that in Russia both the Federal Security Service (FSB) and the Ministry of Defense were the lead agencies for cyberwar activities.

CONCLUSION

It is a far cry from the Peloponnesian War to Stuxnet—in both years and technology—and a student of military affairs in the 21st century might be forgiven for being unable to find any common principles that would make it possible to discuss these two events in the same conversation. Part of the student's perplexity, however, might stem from an undeveloped historical imagination, and here the 21st-century student might have something in common with the greatest leaders of the previous two centuries. For most politicians and generals in the 19th and 20th centuries, war meant the kind of conflict characteristic of European contests from the middle of the 17th century to their own time: state-centred, conducted by increasingly professional armed forces, nominally excluding civilians, and involving well-defined instruments commonly available in developed states. These were wars that began with declarations and ended with armistices or treaties; they might last weeks, months, or even years, but they had definite beginnings and endings.

In the emerging world of the 21st century, it seems more reasonable to reach farther back in time for models of current events. Medieval warfare, after all, could last decades, even centuries. It involved states and trans- or substate organizations—even philanthropic organizations. Religion provided powerful motivation, but so too did state and even personal interest. High politics mixed with banditry, and even the most powerful persons and societies were subject to acts of extraordinary savagery and cruelty. No prudent political leader in the United States could publicly describe that country's war with al-Qaeda as a "crusade," but a thoughtful military historian might point to the parallels.

In such a world the classical paradigm of the 19th and 20th centuries, however modified, still holds some value. If one understands politics broadly enough—as the way in which human societies rule themselves, define and administer justice, and articulate their visions of what is possible and good—war remains very much about politics. The logic of struggle between interacting entities remains. It accounts for the possibility of surprise that forms so large an element in war. The fact that violence, however used, engages the emotions and thereby influences (and sometimes overwhelms) judgment remains true. The advent of weapons that can obliterate cities, and that may be available to small groups of terrorists and not just states, may make the stakes of war in the 21st century even higher, but they were also enormous in the great World Wars of the 20th century.

War is a discipline of thought as well as a practical art. As it has become ever more complex, its dependence on a wide array of allied disciplines has grown. To understand war in the modern world, one must understand something about development economics and bioengineering, as well as precision guidance and computer programming. In war studies more than in other practical fields, there has long existed a craving for simple formulas and aphorisms: "the offense needs an advantage of three to one," for example. Such formulas may never have been terribly useful, but in a new and no less dangerous century they are less helpful than ever in steering citizen and soldier alike through choices that remain as consequential as they have ever been.

armistice A temporary suspension of hostilities by agreement of all warring parties; truce.

attrition A wearing down or weakening of resistance or an enemy force by constant attack.

barrage A heavy onslaught of military fire to stop the advance of enemy troops.

battalion A military unit composed of a headquarters and two or more companies, batteries, or similar units.

blockade The closing off of a certain area by troops to prevent the enemy from entering or exiting.

breech loading Describes firearm that loads at the breech, the part of the firearm at the rear of the barrel.

cadre Key group of military officers necessary to train and establish a new military unit.

carbines A short-barreled lightweight firearm originally used by cavalry.

cavalry Military force mounted on horseback, formerly an important element in the armies of all major powers.

coalition A provisional alliance of diverse parties, persons, or states for cooperative action.

company A unit (as of infantry) consisting usually of a headquarters and two or more platoons.

concertina wire A coiled barbed wire for use as an obstacle.

conscription Required enrollment in the military; the draft.

constable A high officer of a royal court, especially in the Middle Ages.

convoy A group of land vehicles or ocean vessels traveling together.

coup de grace Death blow.

cuirassiers Heavy cavalry.

deterrence The maintenance of military power for the purpose of discouraging attack.

dragoon A member of a European military unit formerly composed of heavily armed mounted troops.

echelon An arrangement of a body of troops with its units each somewhat to the left or right of the one in the rear, like a series of steps.

fascine A bundle of sticks bound together and used for such purposes as filling ditches.

garrison A military post especially a permanent military installation.

halberd A weapon of the 15th and 16th centuries consisting of a battle-ax and pike mounted on a handle about six feet long.

harquebus A matchlock gun invented in the 15th century that was portable but heavy and was usually fired from a support.

hussar A member of any of various European military units originally modeled on the Hungarian light cavalry of the 15th century.

impregnable Strong enough to withstand attack.

incursion A hostile invasion.

infantry Soldiers who fight on foot.

insurrection An act of rising in revolt against civil authority or organized government.

interdiction The steady bombardment of enemy communication lines to delay their advance.

irregulars Free corps, or soldiers who are not members of a regular military force.

matériel Ammunition, weapons, and other military equipment.

nonbelligerent Not participating in a conflict.

ordnance Military supplies, including weapons and ammunition.

platoon A subdivision of a company-sized military unit normally consisting of two or more squads or sections.

poliorcetics The art of both fortification and siege warfare.

reprisal An act of retaliation.

salient An outwardly projecting part of a fortification, trench system, or line of defense.

salvo Simultaneous discharge of artillery or firearms.

scuttled Sunk purposefully by allowing water to pass through the hull of a ship.

Bibliography

Strategy

Edward Mead Earle, Gordon A. Craig, and Felix Gilbert (eds.), *Makers of Modern Strategy: Military Thought from Machiavelli to Hitler* (1941, reissued 1971), remains a most competent anthology on the development of the military mind and art. Its worthy successor is Peter Paret, Gordon A. Craig, and Felix Gilbert (eds.), *Makers of Modern Strategy: From Machiavelli to the Nuclear Age* (1986), which repeats three of the essays and provides more than 20 new ones. To both should be added Williamson Murray, MacGregor Knox, and Alvin Bernstein (eds.), *The Making of Strategy: Rulers, States, and War* (1994), which looks at strategy as a collective rather than an individual function.

Among the classic texts of strategic thought, the two most important are Carl von Clausewitz, *On War*, ed. and trans. by Michael Howard and Peter Paret (1976, reissued 1989); and Sunzi, *Sun Tzu: The Art of War: The First English Translation Incorporating the Recently Discovered Yin-ch'üeh-shan Texts*, trans. by Roger T. Ames (1993). Geoffrey Blainey, *The Causes of War*, 3rd ed. (1988), is the best short modern introduction to the origins of wars. To these one should add the masterly collection of essays by Michael Howard, *The Causes of War* (1984); and, for a study on the relationship between politicians and generals in modern war, Eliot A. Cohen, *Supreme Command: Soldiers, Statesmen, and Leadership in Wartime* (2002).

Thucydides, *The Landmark Thucydides: A Comprehensive Guide to the Peloponnesian War* (1996), is an excellent starting point for exploring the history of military strategy. Hew Strachan, *The First World War*, vol. 1, *To Arms* (2001), begins the monumental three-volume history covering strategy in World War I. For works that explore the full complexity of high command at the top, the reader should turn to broad military histories, such as the extraordinary British official history of strategy during World War II, *Grand Strategy*, 6 vol. in 7 (1956–76), a Her Majesty's Stationery Office publication; or the more limited Maurice Matloff and Edwin M. Snell, *Strategic Planning for Coalition Warfare*, 2 vol. (1953–59, reprinted 1968; vol. 1 also reissued separately, 1999), covering the years 1941–44. There are good accounts of strategic decision making in more recent conflicts in Michael B. Oren, *Six Days of War: June 1967 and the Making of the Modern Middle East* (2002); and Michael R. Gordon and Bernard E. Trainor, *The Generals' War: The Inside Story of the Conflict in the Gulf* (1995).

Finally, the reader should not disregard the works about and by the makers of strategy themselves. In particular, Winston S. Churchill, *Marlborough: His Life and Times*, 6 vol. (1933–38, reissued

2002), is a strategical treatise masquerading as biography. Alan Brooke, Viscount Alanbrooke, *War Diaries 1939–1945*, ed. by Alex Danchev and Daniel Todman (2001), the dyspeptic memoirs of the head of the British army during World War II, gives a useful window into the stresses of war at the top and an unwitting revelation of the difficulty of judging aright in the midst of war. Geoffrey Parker, *The Grand Strategy of Philip II* (1998), demonstrates that biographies from the early modern period and even before also can be studied with profit.

TACTICS

For a general introduction to tactics, see Martin van Creveld, *Technology and War: From 2000 B.C. to the Present* (1989). John Keegan, *The Illustrated Face of Battle: A Study of Agincourt, Waterloo, and the Somme*, rev. ed. (1989), is excellent on the tactics of three critical, widely separated battles; see also Arther Ferrill, *The Origins of War: From the Stone Age to Alexander the Great* (1985), which covers the period indicated while arguing that tactics underwent no basic change from the earliest time to Waterloo.

The best book on tribal warfare remains Harry Holbert Turney-High, *Primitive War: Its Practice and Concepts*, 2nd ed. (1971). Biblical warfare is covered in Yigael Yadin, *The Art of Warfare in Biblical Lands: In the Light of Archaeological Study*, 2 vol. (1963; originally published in Hebrew, 1963). On ancient warfare in general, see Peter Connolly, *Greece and Rome at War* (1981); and Victor Davis Hanson, *The Western Way of War: Infantry Battle in Classical Greece* (1989). Of several excellent books on medieval warfare, J.F. Verbruggen, *The Art of Warfare in Western Europe During the Middle Ages: From the Eighth Century to 1340* (1977; originally published in Dutch, 1954), is perhaps the strongest on tactics.

The most expert contemporary work on early modern warfare is undoubtedly Geoffrey Parker, *The Military Revolution: Military Innovation and the Rise of the West, 1500–1800* (1988). The early 18th century is covered in David Chandler, *The Art of Warfare in the Age of Marlborough* (1976), excellently researched and well written. For subsequent developments in the same century, see Christopher Duffy, *The Military Experience in the Age of Reason* (1988); as well as Robert S. Quimby, *The Background of Napoleonic Warfare: The Theory of Military Tactics in Eighteenth-Century France* (1957, reprinted 1968), a meticulous inquiry into tactics before and during the French Revolution.

Larry H. Addington, *The Patterns of War Since the Eighteenth Century* (1984); and Hew Strachan, *European Armies and the Conduct of War* (1983), are good general accounts. Two older works that can still be read with profit are Theodore Ropp, *War in the Modern World* (1959, reprinted 1981); and J.F.C. Fuller, *The Conduct of War, 1789–1961: A Study of the Impact of the French, Industrial, and Russian Revolutions on War and Its Conduct* (1961, reprinted 1981). William McElwee, *The Art of War: Waterloo to Mons* (1974), is probably the best of many works on 19th-century warfare.

For the tactics of World War I in general, see Tony Ashworth, *Trench Warfare, 1914–1918: The Live and Let Live System* (1980); on the offensive tactics developed by the Germans, Timothy T. Lupfer, *The Dynamics of Doctrine: The Changes in German Tactical Doctrine During the First World War* (1981), is excellent. For World War II, B.H. Liddell Hart, *History of the Second World War* (1971, reissued 1982), though flawed on some counts, remains the single most comprehensive operational history. On armoured warfare, see F.W. von Mellenthin, *Panzer Battles: A Study of the Employment of Armor in the Second World War*, trans. from German (1956, reissued 1982); and Charles Messenger, *The Blitzkrieg Story* (1976); for the ways of countering it, see John Weeks, *Men Against Tanks: A History of Anti-Tank Warfare* (1975).

The story of the Korean War is ably told in Callum A. MacDonald, *Korea, the War Before Vietnam* (1987); that of the Arab-Israeli Wars, in Trevor N. Dupuy, *Elusive Victory: The Arab-Israeli Wars, 1947–1974* (1978, reissued 1984). For the Vietnam War, see Andrew F. Krepinevich, Jr., *The Army and Vietnam* (1986). Michael Carver, *War Since 1945* (1980, reissued 1990), provides an excellent general overview.

Logistics

Martin van Creveld, *Supplying War: Logistics from Wallenstein to Patton*, 2nd ed. (2004), is an insightful groundbreaking history of logistics. Classic studies of the subject include George Cyrus Thorpe, *George C. Thorpe's Pure Logistics: The Science of War Preparation*, new ed., with an introduction by Stanley L. Falk (1986); G.C. Shaw, *Supply in Modern War* (1938), mainly on subsistence; S.L.A. Marshall, *The Soldier's Load and the Mobility of a Nation* (1950, reprinted 1980); and Henry E. Eccles, *Logistics in the National Defense* (1959, reprinted 1981), with emphasis on theory.

The 18th-century logistics systems are examined in Lee Kennett, *The French Armies in the Seven Years' War: A Study in Military Organization and Administration* (1967, reprinted 1986); and Erna Risch, *Supplying Washington's Army* (1981). The U.S. Army experience is surveyed in James A. Huston, *The Sinews of War: Army Logistics 1775–1953* (1966, reprinted 1988); Edwin A. Pratt, *The Rise of Rail-Power in War and Conquest, 1833–1914* (1916), an old but still useful survey; Robert Greenhalgh Albion and Jennie Barnes Pope, *Sea Lanes in Wartime: The American Experience 1775–1945*, 2nd ed. (1968); and C.B.A. Behrens, *Merchant Shipping and the Demands of War*, rev. ed. (1978), on the overseas supply.

Roland G. Ruppenthal, *Logistical Support of the Armies*, 2 vol. (1953–59, reprinted 1985–87); and Richard M. Leighton and Robert W. Coakley, *Global Logistics and Strategy*, 2 vol. (1955–68), provide the U.S. Army's official history of logistics in World War II, in the European theatre and in the framework of coalition strategy, respectively. R. Elberton Smith, *The Army and Economic Mobilization* (1959, reprinted 1985), analyzes the U.S. Army's World War II economic mobilization.

Charles J. Hitch and Roland N. McKean, *The Economics of Defense in the Nuclear Age* (1960, reissued 1978), is the "bible" of the managerial reforms in the U.S. Defense Department; and Neville Brown, *Strategic Mobility* (1963), explores a facet of post-World War II international strategy and logistics.

Guerrilla Warfare

Robert B. Asprey, *War in the Shadows: The Guerrilla in History* (1975, reissued 2002), surveys guerrilla and counter-guerrilla warfare from its origin to the 1990s. Other surveys can be found in Richard L. Clutterbuck, *Terrorism and Guerrilla Warfare* (1990); and Ian F.W. Beckett, *Modern Insurgencies and Counter-Insurgencies: Guerrillas and Their Opposites Since 1750* (2002). David Kilcullen, *The Accidental Guerrilla: Fighting Small Wars in the Midst of a Big One* (2009), argues that counterinsurgency efforts are doomed to failure if they focus solely on ideological conflict and do not take into account the local and cultural factors that drive insurgents to take up arms.

INDEX

D

E

F

G

H

I

J

K

L

Q

R

S

T